Advances in Anatomy, Embryology and Cell Biology
Ergebnisse der Anatomie und Entwicklungsgeschichte
Revues d'anatomie et de morphologie expérimentale

Springer-Verlag Berlin Heidelberg New York

This journal publishes reviews and critical articles covering the entire field of normal anatomy (cytology, histology, cyto- and histochemistry, electron microscopy, macroscopy, experimental morphology and embryology and comparative anatomy). Papers dealing with anthropology and clinical morphology will also be accepted with the aim of encouraging co-operation between anatomy and related disciplines.

Papers, which may be in English, French or German, are normally commissioned, but original papers and communications may be submitted and will be considered so long as they deal with a subject comprehensively and meet the requirements of the Ergebnisse.

For speed of publication and breadth of distribution, this journal appears in single issues which can be purchased separately; 6 issues constitute one volume.

It is a fundamental condition that submitted manuscripts have not been, and will not simultaneously be submitted or published elsewhere. With the acceptance of a manuscript for publication, the publishers acquire full and exclusive copyright for all languages and countries.

25 copies of each paper are supplied free of charge.

Les résultats publient des sommaires et des articles critiques concernant l'ensemble du domaine de l'anatomie normale (cytologie, histologie, cyto et histochimie, microscopie électronique, macroscopie, morphologie expérimentale, embryologie et anatomie comparée. Seront publiés en outre les articles traitant de l'anthropologie et de la morphologie clinique, en vue d'encourager la collaboration entre l'anatomie et les disciplines voisines.

Seront publiés en priorité les articles expressément demandés nous tiendrons toutefois compte des articles qui nous seront envoyés dans la mesure où ils traitent d'un sujet dans son ensemble et correspondent aux standards des «Résultats». Les publications seront faites en langues anglaise, allemande et française.

Dans l'intérêt d'une publication rapide et d'une large diffusion les travaux publiés paraitront dans des cahiers individuels, diffusés séparément: 6 cahiers forment un volume.

En principe, seuls les manuscrits qui n'ont encore été publiés ni dans le pays d'origine ni à l'étranger peuvent nous être soumis. L'auteur d'engage en outre à ne pas les publier ailleurs ultérieurement.

Les auteurs recevront 25 exemplaires gratuits de leur publication.

Die Ergebnisse dienen der Veröffentlichung zusammenfassender und kritischer Artikel aus dem Gesamtgebiet der normalen Anatomie (Cytologie, Histologie Cyto- und Histochemie, Elektronenmikroskopie, Makroskopie, experimentelle Morphologie und Embryologie und vergleichende Anatomie). Aufgenommen werden ferner Arbeiten anthropologischen und morphologisch-klinischen Inhaltes, mit dem Ziel, die Zusammenarbeit zwischen Anatomie und Nachbardisziplinen zu fördern.

Zur Veröffentlichung gelangen in erster Linie angeforderte Manuskripte, jedoch werden auch eingesandte Arbeiten und Originalmitteilungen berücksichtigt, sofern sie ein Gebiet umfassend abhandeln und den Anforderungen der „Ergebnisse" genügen. Die Veröffentlichungen erfolgen in englischer, deutscher und französischer Sprache.

Die Arbeiten erscheinen im Interesse einer raschen Veröffentlichung und einer weiten Verbreitung als einzeln berechnete Hefte; je 6 Hefte bilden einen Band.

Grundsätzlich dürfen nur Arbeiten eingesandt werden, die nicht gleichzeitig an anderer Stelle zur Veröffentlichung eingereicht oder bereits veröffentlicht worden sind. Der Autor verpflichtet sich, seinen Beitrag auch nachträglich nicht an anderer Stelle zu publizieren.

Die Mitarbeiter erhalten von ihren Arbeiten zusammen 25 Freiexemplare.

Manuscripts should be addressed to/Envoyer les manuscrits à/Manuskripte sind zu senden an:

Prof. Dr. A. BRODAL, Universitetet i Oslo, Anatomisk Institutt, Karl Johans Gate 47 (Domus Media), Oslo 1/Norwegen

Prof. W. HILD, Department of Anatomy. The University of Texas Medical Branch, Galveston, Texas 77550 (USA)

Prof. Dr. J. van LIMBORGH, Universiteit van Amsterdam, Anatomisch-Embryologisch Laboratorium, Amsterdam-O/Holland, Mauritskade 61

Prof. Dr. R. ORTMANN, Anatomisches Institut der Universität, D-5000 Köln-Lindenthal, Lindenburg

Prof. Dr. T. H. SCHIEBLER, Anatomisches Institut der Universität, Koellikerstraße 6, D-8700 Würzburg

Prof. Dr. G. TÖNDURY, Direktion der Anatomie, Gloriastraße 19, CH-8006 Zürich

Prof. Dr. E. WOLFF, Collège de France, Laboratoire d'Embryologie Expérimentale, 49 bis Avenue de la belle Gabrielle, Nogent-sur-Marne 94/France

Advances in Anatomy, Embryology and Cell Biology
Ergebnisse der Anatomie und Entwicklungsgeschichte
Revues d'anatomie et de morphologie expérimentale

50·4

Editors
A. Brodal, Oslo · W. Hild, Galveston · J. van Limborgh, Amsterdam
R. Ortmann, Köln · T. H. Schiebler, Würzburg · G. Töndury, Zürich · E. Wolff, Paris

Willi A. Ribi

The Neurons of the First Optic Ganglion of the Bee *(Apis mellifera)*

With 21 Figures

Springer-Verlag Berlin Heidelberg New York 1975

Dr. Willi A. Ribi
Zoologisch-Vergl. Anatomisches
Institut der Universität Zürich
Künstlergasse 16
CH-8006 Zürich/Schweiz

Present address:
Department of Neurobiology,
Research School of Biological Sciences,
Australian National University,
Canberra, A.C.T. 2600, Australia

Library of Congress Cataloging in Publications Data

Ribi, W   A   1943-
The neurons of the first optic ganglion of the bee (Apis mellifera).

(Advances in anatomy, embryology and cell biology; 50/4)
Biography: p.
Includes index.
1. Bees—Anatomy. 2. Optic lobes. I. Title. II. Series. [DNLM: 1. Bees—Anatomy and histology. 2. Optic chiasm—Anatomy and histology. W1 AD433K v. 50 fasc. 4/QX565 R485n]
QL801.E67 vol. 50, fasc. 4 [QL568.A6] 574.4'08s

ISBN 978-3-540-07096-2   ISBN 978-3-642-50057-2 (eBook)
DOI 10.1007/978-3-642-50057-2

This work is subject to copyright. All rights are reserved, whether the whole or part of the material is concerned, specifically those of translation, reprinting, re-use of illustrations, broadcasting, reproduction by photocopying machine or similar means, and storage in data banks.

Under § 54 of the German Copyright Law where copies are made for other than private use, a fee is payable to the publisher, the amount of the fee to be determined by agreement with the publisher. © by Springer-Verlag Berlin-Heidelberg 1975.

The use of general descriptive names, trade names, trade marks, etc. in this publication, even if the former are not especially identified, is not to be taken as a sign that such names, as understood by the Trade Marks and Merchandise Marks Act, may accordingly be used freely by anyone.

Printed by H. Stürtz AG, Universitätsdruckerei, 8700 Würzburg, Germany.

Contents

Symbols	6
I. *Introduction*	7
II. *Material and Methods*	7
1. Reduced silver impregnation	8
2. Methylene blue staining	8
3. Golgi's selective silver impregnation	9
4. Technical equipment	9
III. *Results*	10
1. Morphology of the lamina	10
Fenestrated layer (FL)	10
Monopolar cell body layer (CBL)	13
External plexiform layer (EPL)	14
2. Localization of cell bodies in the peripheral visual system	14
3. Retina- lamina and lamina- medulla projections	14
4. Lamina fibres	17
Retinula-cell-axons (R-fibres)	17
Monopolar cells (L-fibres)	21
Tangential fibres (T-fibres)	25
Centrifugal fibres (C-fibres)	28
Amacrines (am)	28
Inserta sedis (is)	28
5. Outer chiasma	28
6. Medulla	28
Terminals of long visual fibres (lvf)	28
Monopolar-cell Terminal (L-fibres)	28
IV. *Discussion*	32
Summary	39
1. General anatomical features	39
2. Retinula- (svf and lvf) and lamina-fibers (L-, am-, T-, C-,)	39
3. Discussion of the connection pattern	40
Acknowledgements	41
References	41
Subject Index	43

Symbols

A	axon-bundle
am	amacrine cell
am i	amacrine of inner stratum
am b	amacrine bistratified
BM	basal membrane
Ca	cartridge
CBL	cell body layer
C-fibre	centrifugal fibre
EPL	external plexiform layer
FL	fenestrated layer
ICh	inner Chiasma
is	inserta sedis, unknown origin
L	lamina ganglionaris
L-fibre	monopolar cell of lamina
Lo	lobula
lvf	long visual fibre
M	medulla externa
OCh	outer Chiasma
PC	pseudocartridge, axonbundle of one ommatidium
R	retina
R-fibre	retinula cell axon
R(d)	deep retinula cell
R(s)	shallow retinula cell
svf	short visual fibre
T-fibre	tangential fibre
T(nf)	narrow field T-cell
T(wf)	wide field T-cell

I. Introduction

The visual system of insects has attracted histologists for a long time. We have detailed histological studies of the visual systems of Diptera, Hymenoptera and Odonata dating from the last century: Leydig's (1864) study on optical ganglia of insects, Ciaccio's (1876) on the fine structure of the first ganglion in the mosquitos and Hickson's (1885) giving for the first time an exact description of the three optical ganglia of *Musca*. From 1896 several papers appeared using neuro-histological methods, mainly Golgi techniques and methylene blue staining. In 1896 Kenyon published his work on the bee-brain and in 1897 described the relationships of neurons in the optic ganglia with a modified Golgi method. Another work, by Jonescu (1909), should be mentioned: "Vergleichende Untersuchungen am Gehirn der Honigbiene". In the same year Cajal's findings on the optic ganglia of the fly were published; then in 1915 Cajal and Sanchez wrote a definitive monograph on the neural elements and their connections in the same animal that remained the main reference in this field for decades. In both works Golgi techniques were used.

Since then there have been only a few new publications on the subject: (Gribakin, 1967; Perrelet and Baumann, 1969a, b; Perrelet, 1970; Varela and Porter, 1969; Skrzipek and Skrzipek, 1971, 1973; Grundler, 1972: Snyder, Menzel and Laughlin, 1973). They deal mainly with the receptors of the retina.

There is only one publication on the structure of the lamina (Strausfeld, 1970) where some fibre types are described in Golgi preparations. Varela (1970) mentions some ultrastructural features of the lamina, but they alone are not enough to allow an understanding of the whole lamina. For that reason a light-microscope analysis of the neuroanatomy of the first optic ganglion and of its neighbouring structures was undertaken. These results may be important for further behavioural and electrophysiological research.

II. Material and Methods

Worker bees and drones of the species *Apis mellifera* (sub-species mellifera) were used. They came from populations kept at the Zoological Institute of the University of Zurich and at the Max-Planck-Institute for Biological Cybernetics of Tubingen, Germany. The drones were taken from the hive just after eclosion; workers without particular regard to age.

To understand neural functions and to interpret the relation between their elements it is necessary to know the position of the cell bodies, the structure of each neuron and to map the whole network. There being no single method available to achieve all these goals, three techniques were used some of which were especially adapted for the bee: reduced silver impregnation for the staining of all neurons, Golgi technique for selective staining of neurons, and methylene blue staining to localize cell bodies.

1. Reduced Silver Staining

Bodian's (1936), Holme's (1943), Blest's (1961), Rowell's (1963) and Weiss's (1972) reduced silver impregnation, although successfully used on Dipterans, did not work on the bee. Thus we developed three impregnation procedures (Ribi, 1974) related to Cajal's (1915) silver-formaldehyde- and Holme's (1943) technique. These procedures were very satisfactory on the bee and also produced beautiful staining in the Dipteran (*Phormia regina*) and the Orthopteran (*Locusta migratoria*).

(a) Worker-bees and drones were paralysed by cooling them to 4° C. Usually they were motionless after 5–7 minutes.

(b) Next the mandibles were removed at the level of the epistomal suture; cranial and caudal parts of the head-capsule were cut off. This resulted in a preparation where the brain was surrounded by the dorsal and lateral skeleton only, so that the fixative could enter the neural tissue easily.

(c) The following fixatives, listed in order of increasing effectiveness were used: Carnoy's solution, formaldehyde, Duboscq-Brasil-solution, glutaraldehyde, Bouin's solution (from Romeis, 1948) and glutaraldehyde-Bouin combined. All solutions were mixed fresh for each series.

The glutaraldehyde-Bouin method had the advantage of staining not only the fibres but also the cell bodies of the optic ganglia. We will therefore give a more detailed account of this procedure.

The surgically prepared bee-heads were prefixed in a phosphate-buffered glutaraldehyde-paraformaldehyde solution (Karnovsky, 1965). After four hours the heads were washed several times for a total of one hour in phosphate buffer and then transferred to Bouin's solution for 8–12 hours. This solution was removed by 70% ethanol (12 hours, several changes).

(d) After dehydration with ethanol the brains were immersed in methylbenzoate for 18 hours (3 changes), in benzene ($3 \times 1/2$ hour) and finally in a benzene-Histoplast-mixture (1:1, 1 hour). They were embedded in pure Histoplast (12 hours).

(e) The paraffin blocks were cut on a sledge microtome, and sections of 10–15 µm were made as thinner sections do not impregnate well.

(f) To prevent the sections from falling off the slides, these were cleaned in a 1:1 mixture of ethanol and ether for two days and dried in an oven. After drying the slides were immersed in a solution of 3 gm of white gelatine and 0.5 gm of chrom-alum in 1 litre water. The gelatine was diluted by heating, the chrom-alum was added after cooling. The slides are again dried in an oven at 40° C. The three following reduced silver impregnations were used:

Method A: After removing the paraffin from the sections, they were kept for three hours in a 20% $AgNO_3$-solution at 20° C in the dark, washed in distilled water and immersed for 3 min in a hydroquinone-solution (1 gm hydroquinone, 10 gm sodium sulfate in 100 ml of dist. water), washed again shortly in distilled water and finally laid for 10 minutes in a solution of 1% gold-chloride. Before and after developing the sections in 2% oxalic acid (10 min) they were washed 3 times for 3 minutes in distilled water. The fixation occured in a 5% solution of sodium thiosulfate (10 min) after which they were embedded in Permount (Fisher).

Method B: The paraffin-free sections were incubated for 3 hours at room temperature in a 20% $AgNO_3$-solution (in the dark), then washed (distilled water, 3 times 1 min) and intensified in the following solution: 55 ml boric acid (6, 2 gm in 500 ml distilled water), 45 ml borax (9, 5 gm in 500 ml distilled water), 2 ml of a 1.5% $AgNO_3$-solution, 5 ml of a 10% pyridine-solution and 500 ml distilled water for 48 hours in the dark at 37° C. The sections were then washed and reduced in hydroquinone and sodium sulfate. The remaining steps are the same as for method A.

Method C: This was identical to method B except for one point: instead of pyridine, formol was used (same concentration and volume).

2. Methylene blue Staining

Semi-thin sections coloured with this method showed clearly the localization and number of nerve-, glial-, and tracheoblast cell bodies which appear dark blue on a more or less colourless background.

(a) The preparation of the bee-heads was the same as in the previous method.

(b) For fixation Carnoy's solution was used as it travels rapidly through the tissue. This is a result firstly of the dissociation of H_3O^+-ions in the water-rich tissue caused by the acetic acid in the solution, and secondly the chloroform dilutes fats and lipids. Both these processes allow a faster absorption of alcohol.

(c) Fixation time was 5 hours at most; longer times resulted in tissue-shrinking and extreme hardening.

(d) Fixed material was dehydrated in ethanol; propylene-oxide (2 × 15 min) then left for 48 hours in a mixture of propylene-oxide and Araldite (Ducurpan, Fluka) (1:1) and finally embedded in Araldite as used for electron microscopy.

(e) The semi-thin sections were cut on an ultramicrotome with glass knives, transferred to a distilled water covered slide and left to dry at 60° C.

(f) The staining solution had three components: solution A: 1% Azur A (Serva); solution B: 1% borax; solution C: 1% methylene blue · HCl (Serva). The substances were dissolved in distilled water. Solution D is 1% of solution C and 99% solution B. A and D mixed 1:1 are the staining solution E.

The filtered solution E was applied one drop section and continuously heated until a golden ring formed around each drop. The rest of the dye was then washed off with distilled water, the slides left to dry on a hot plate. Over-stained sections were differentiated with 80% ethanol with a few drops of ice acid. The dry sections were embedded in Permount (Fisher).

3. Golgi's Selective Silver Impregnation

This method dating from the turn of the century allows a selective staining of single neurons which take on a dark-brown, dark-red or black colour due to the precipitation of silver. Why only certain neurons are stained and not others is still unknown. The background is a pale yellow. If the sections are 80–100 μm thick it is possible to trace whole neurons with all their branches so that a catalogue of different neuron-types and their possible connections may be compiled.

The method used in this work was Collonier's (1964) variation of the Golgi technique. Other variations for insect-tissue have been devised by Strausfeld (1971).

(a) Anaesthetised bees were prepared in such a way that the brain-tissue was in direct contact with the fixative. This facilitates the perfusion of peripheral brain-parts. Small pieces of cornea were cut from both compound eyes.

(b) The fixative was made of 4 parts 2.5% potassium bichromate and 1 part of 25% glutaraldehyde (Merck or Fisher). The fixation time was 6 days at room temperature.

(c) The fixed heads were washed repeatedly in a 0.75% $AgNO_3$-solution for several times until no more red precipitate could be discerned. The staining procedure lasted for at least 6 days, longer staining resulted in higher impregnation.

(d) From the silver solution the heads were transferred to 30%, 50%, 70%, 80%, 95% ethanol (15 min each), then to 100% ethanol (2 × 10 min each) and finally to propylene-oxide (2 × 20 min).

(e) The embedding was in a specially soft Araldite (Durcupan, Fluka) or Epon mixture for light-microscopy in two steps. The first solution was Araldite-propylene-oxide (1:1) and the samples were left overnight and then transferred into pure Araldite for 4 hours. The polymerisation occured in a fresh Araldite mixture at 60° C and was complete after 24–36 hours.

(f) The blocks were cut on a sledge microtome with Schick-blades in 30–120 μm thick sections. Before embedding in Permount on the slides, the sections were flattened with a drop of xylene. Sections were cut in the horizontal tangential and frontal planes (Fig. 1).

Six hundred Golgi preparations, three hundred reduced silver impregnations and a hundred section-series were used for these investigations.

4. Technical Equipment

The light-microscopy were done with a Leitz-Orthoplan large-field microscope and the corresponding camera. The following objectives were used: Neofluar 100 × oil, Planapochromat 40 × oil, Planachromat 25 × and 16 ×, Neofluar 6.3 × (all from Zeiss); for interference microscopy: the T-equipment (Leitz) and corresponding objectives 100 ×/1.3 oil; 40 ×/0.65

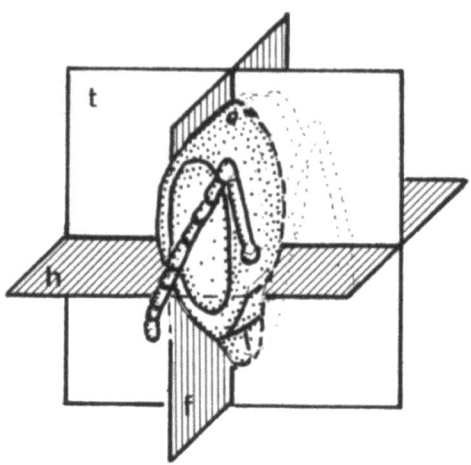

Fig. 1. The three main section-planes of the visual system of the bee: *f* frontal, *h* horizontal, *t* tangential

and 25 × 0.50. To achieve a better contrast on orange Wratten-filter (Nr. 16) and an interference (556 nm)-filter were inserted in the light path (selective and reduced silver-impregnations).

The investigation of the fibres of the lamina and the medulla in semi-thin sections was done in phase-contrast after the method of Zernicke. The films were from ADOX (KB 14, 35 mm, no longer commercially available), the film developer was Neofin-red. Semi-thin sections (0, 5–2 µm) were cut on a Reichert OmU3 ultra-microtome, thick sections (30–120 µm) for Golgi on a Jung-sledge microtome. Normal microtome knives were not suitable for the hard Araldite blocks, Schick-blades gave the best results. Drawings were done with a camera lucida (Leitz) and a graduated slide 10 × 10 × 0.2, three-dimensional reconstructions with a Perspectomat P-40 (Forster, Schaffhausen), and measurements with a micrometer (Leitz, 2 mm in 200).

III. Results

1. Morphology of the Lamina

The lamina ganglionaris or lamina is the outermost ganglion in the visual system of insects (Fig. 2). It lies between the retina distally and the outer or intermediate chiasma proximally. Proximal to the outer chiasma lies the medulla (medulla externa), then the inner chiasma and the lobula (medulla interna). In Hymenoptera there is no lobula plate. The lamina can be divided roughly into a cell body- and a fibre layer (CBL, EPL) (Fig. 3). The CBL consists of a fenestrated zone (FL) and a monopolar cell body layer (MCBL).

The different terminal depths of the short retinula cell axons and a dense horizontal fibre net at the proximal end of the lamina divide the external plexiform layer (EPL) into three parts: A, B and C).

Fenestrated Layer (FL). A dense net of tracheae runs through the neural tissue under the basal membrane (BM) and disrupts the hexagonal pattern of the pseudocartridges, the axon bundles of the ommatidia, before and after the BM. The tracheae (diameter up to 30 µm) make their way between the pseudocartridges

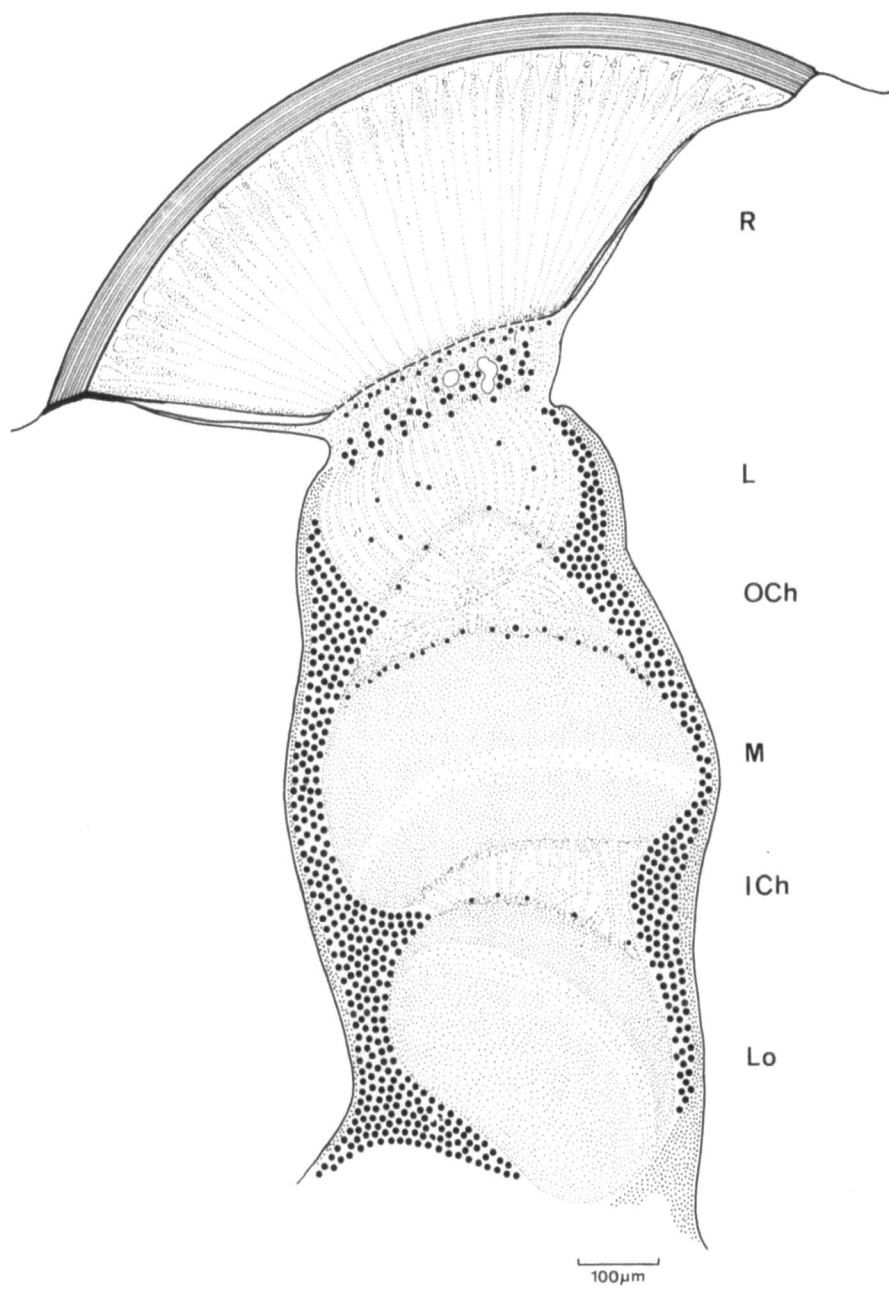

Fig. 2. The right tractus opticus drawn from methylene blue stained horizontal section. Den cell body layers surround the neuropil of the ganglia. *ICh* inner chiasma, *L* lamina, *Lo* lobula, *M* medulla, *OCh* outer (intermediate) chiasma, *R* retina

(Fig. 3). But after having passed the fenestrated layer we find the axon bundles in a new hexagonal pattern in which the x-, y- and z-axes (Braitenberg, 1967) can again be found. A net of fine tracheoles runs through the retina (Fig. 5)

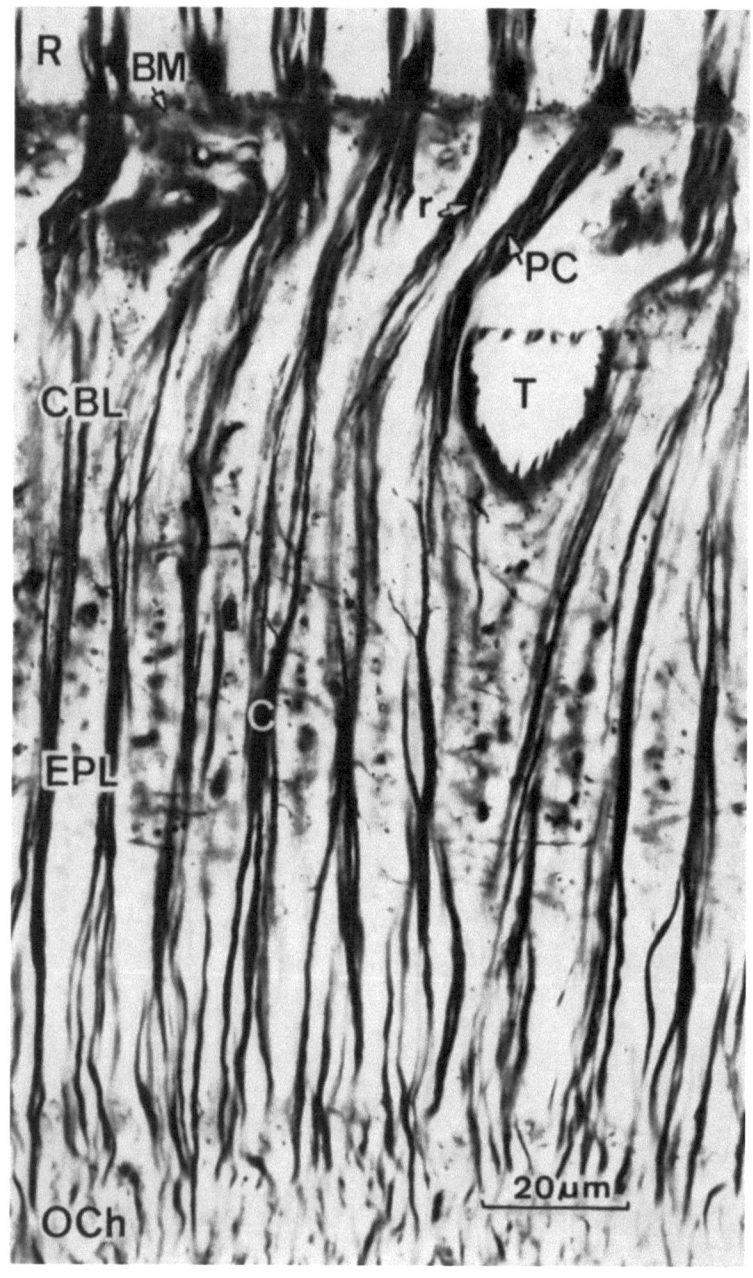

Fig. 3. Frontal section of the lamina (reduced silver impregnation). *BM* basal membrane, *C* cartridge, *CBL* cell body layer, *EPL* external plexiform layer, *OCh* outer chiasma, *PC* pseudocartridge, *R* retina, *T* trachea, *r* twist of the retinula cell axon bundles

and the lamina. Tracheae and tracheoles can easily be distinguished by the fact that the walls of the first are strengthened by a chitinous spiral, whereas the latter have smooth walls.

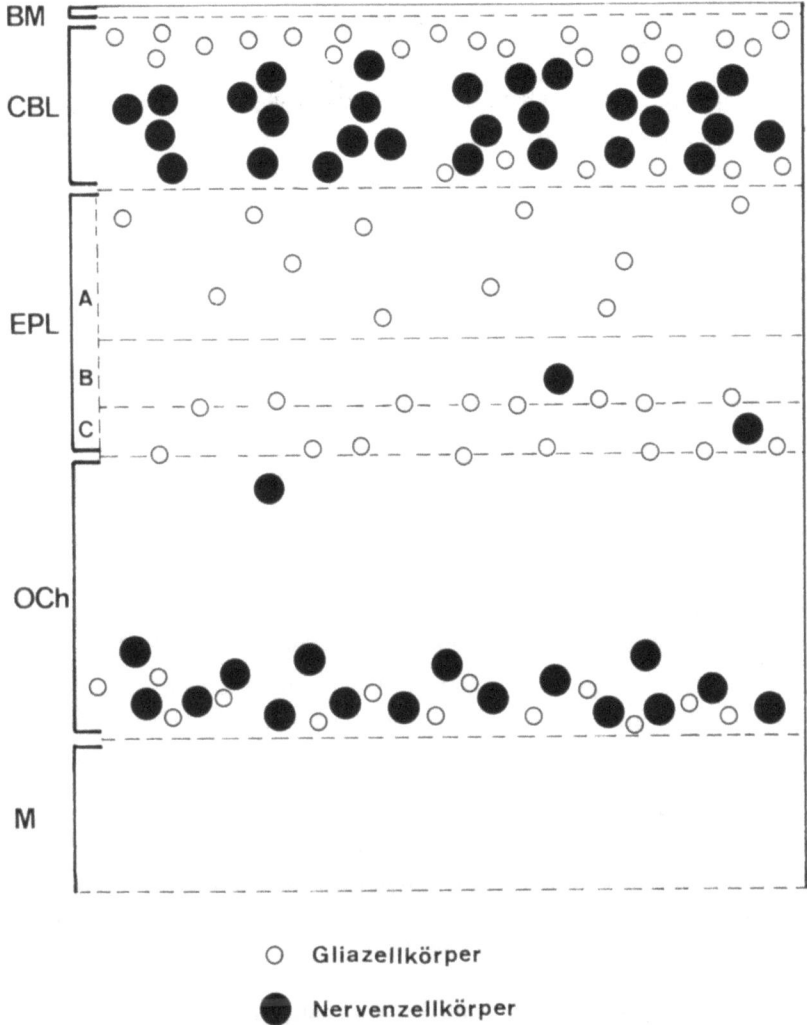

Fig. 4. Distribution of cell bodies in the lamina and outer chiasma from series of methylene blue stained semi-thin sections. Small circles symbolize glia cell bodies lying mostly under the basal membrane (*BM*) in the external plexiform layer (*EPL*) and in the outer chiasma (*OCh*). They are easily distinguishable from larger round monopolar cell bodies. The black dots symbolize the cell bodies of interneurons mostly localized in the cell body layer (*CBL*). The few nerve cell bodies in the outer plexiform layer (*EPL*) probably belong to amacrines. Tangential, transmedullar and medullar fibres have their cell bodies in the proximal part of the chiasma and on the surface of the medulla (*M*)

Monopolar Cell Body Layer (CBL). This zone is sharply defined by the localization of the cell bodies. Groups of ten or more cell bodies are arranged in groups like bunches of grapes to the side of the passing pseudocartridges and tracheae (Fig. 4). Cells of one such group do not belong to any definite cartridge; their cell body diameters range (independent of their location in the lamina) from 6.5 to 10 µm.

The monopolar cell bodies are usually round in methylene blue stained sections, whereas selective silver impregnation produces a variety of shapes which, however, are not correlated with cell types (Figs. 10, 11).

Other cell bodies of the CBL can be distinguished as they are smaller and mainly oblong. They probably belong to glial cells or tracheoblasts.

External Plexiform Layer (EPL). The short R-fibres, R(s)1 and R(s)2, end in EPL(A), their terminals being either simple or forked. Both types have collaterals in this layer. R(d)-fibres reach the EPL(B) as 3 to 4-part tassels.

Branches of the L-fibres may be uni- or bilateral or radially arranged, but each branch type lies in a very definite layer, A, B, C or in a combination of these (Fig. 11).

EPL(C), the most proximal layer is easily recognized by its elements which are oriented perpendicularly to the cartridge- fibers and by its much denser structure as can be seen in interference and Golgi preparations.

The horizontal elements of the EPL consist of the quite long branches of the L-neurons, especially of L(4), the amacrine-fibres running through extensive portions of EPL(C) and some of the centrifugal fibres (Figs. 11 and 13).

2. Localization of Cell Bodies in the Peripheral Visual System

Neurons, glial cells and tracheoblasts of the bee visual system all have different cell body shapes which can be distinguished in methylene blue stained semi-thin sections.

Glial cells have oblong, sometimes slightly curved, cell bodies 3–5 µm long. They are often found in layers directly under the BM, at the distal and proximal end of the EPL or at the surface of the medulla. In the CBL, EPL(A) or (B) they are scattered and none are to be found in the chiasma (Fig. 4). Approximately ten cell bodies of monopolar cells lie in single groups oriented perpendicularly to the BM over the whole width of the lamina (Fig. 4). Their size and shape is characteristic.

A few cell bodies can be found in the external plexiform layer (EPL); they belong to amacrine cells whose axons reach across the EPL(C). The accumulation of cell bodies at the periphery of the medulla might be mostly from tangential fibres as may be inferred from Golgi-stained sections.

The cell bodies of centrifugal fibres have yet to be positively localized, but it is inferred that they might lie in the proximal part of the medulla or even further centrally.

3. Retina-lamina and Lamina-medulla Projections

Methylene blue stained semi-thin sections of the proximal third of the retina show 9 receptor cells per ommatidium, six larger ones and three smaller ones. Sections of the retinula-cell axon bundles also show that of the 6 thicker ones two have a diameter of 2–3 µm and four a diameter of 1.0–1.8 µm. The three thin axons lie in the centre of the ring made of the 6 larger ones and have a fibre cross-section of 0.5–0.8 µm.

In the retina of the bee the microvilli long-axes of corresponding retinula cells in neighbouring fused rhabdoms are not as strictly parallel in orientation as is the

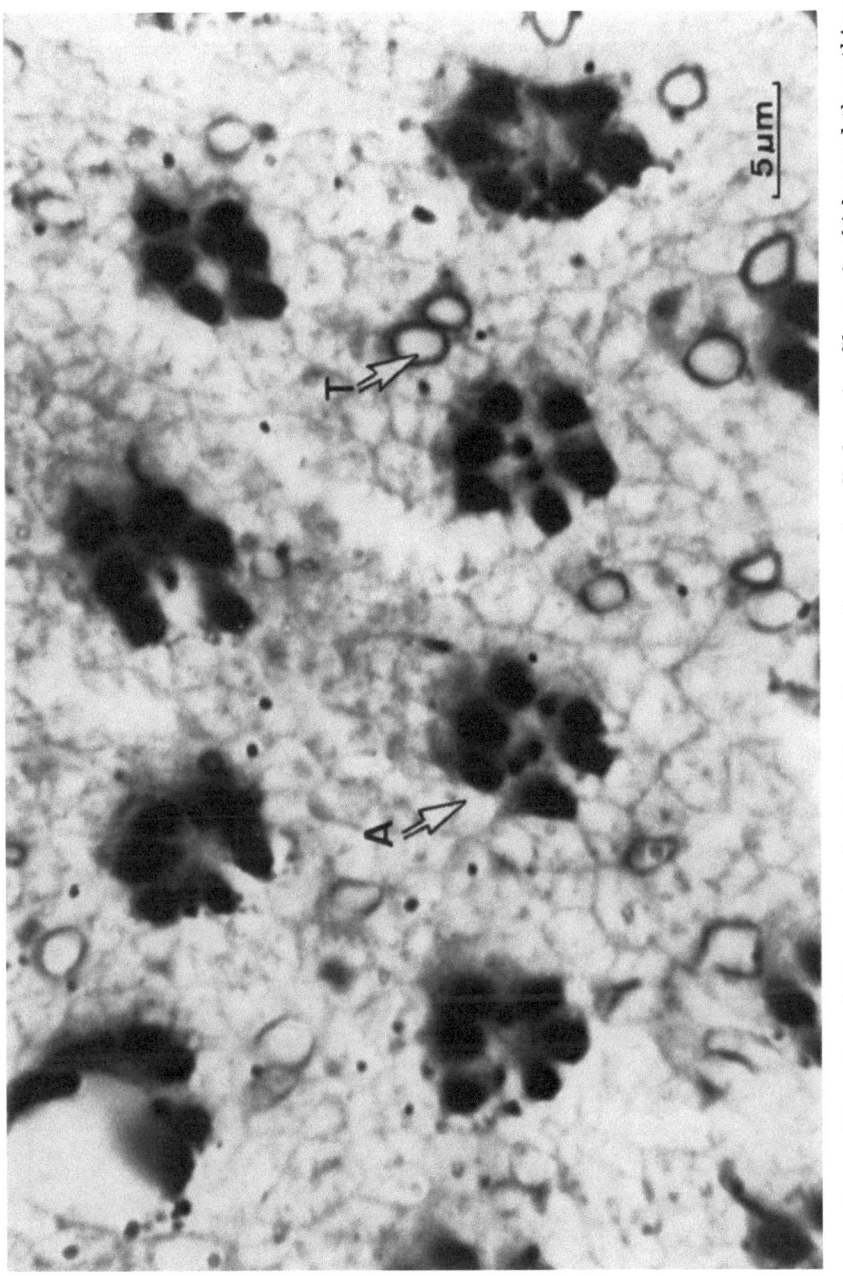

Fig. 5. Tangential section of the proximal end of the retina; each axon bundle has nine fibres: six thicker and three thinner ones. *A* axon bundle, *T* tracheol

case in the open rhabdom of the fly. The internal configuration of axon-bundles however is oriented in a more regular fashion than in the fly (Figs. 5, 6), so much so that the two thickest fibres may be used as a marker. The configuration of each axon bundle remains the same all the way to the first synaptic region although the whole bundle undergoes a torsion of 180° either clockwise or anti-clockwise (Fig. 8). Just above the BM the axons of each pseudocartridge are

15

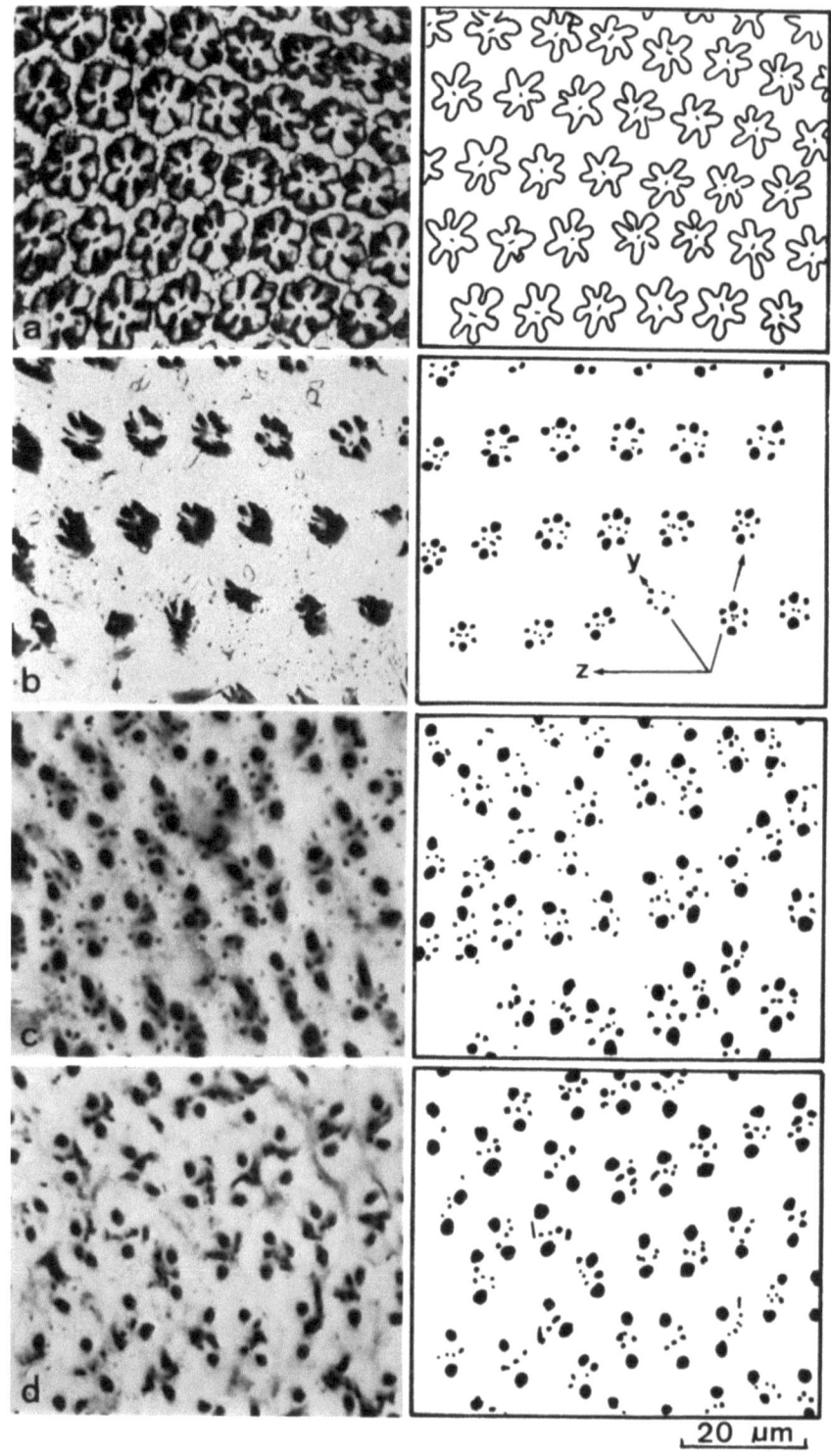

Fig. 6

separated by thick pigment. The BM is the border between the receptors and the nervous tissue (Fig. 3). After the torsion of the axon bundles is complete, the bundles come into contact with a dense net of tangential fibres. These maintain the hexagonal pattern of the bundles and lie between them in such a way that each bundle is separated by every other bundle (Fig. 7).

In Hymenopterans the fibres of one ommatidium do not spread to many cartridges as is the case in Dipterans. At the level of the outer neuropil (EPL(A)) the thin L-fibres join the 9 R-axons and form together a cartridge. The three long visual fibres (lvf) run through the lamina and end in the medulla (Fig. 8). Series of tangential sections (reduced silver impregnation) suggest 3 L-fibres at the level of the R-fibre terminals. In the EPL(C) there are 6 different fibres per bundle. Three are lvf's with a diameter 0.5–0.8 µm, one has a diameter of 3–4 µm and two of 2–3 µm. The latter three might be considered as different L-fibre-types.

Thus there is a 6-fibre bundle running from each cartridge through the outer chiasma to the medulla or second synaptic region. These bundles are separated from each other by neuroglial cells originating from the region where the lamina and the chiasma join. The crossing of the fibres in the horizontal plane occur in such a way that the cartridge pattern of the lamina is projected mirror-like onto the medulla (Fig. 12). Frontal sections show a torsion of the bundles around their own axes. As is the case in the lamina there is a net of tangential fibres at the surface of the medulla such that the incoming bundles form a hexagonal array (Fig. 17). There are six different layers in the medulla according to reduced silver impregnations. Long visual- and L-fibres end in the two first layers (Fig. 16).

4. Lamina Fibres

Retinula-cell Axons (R-fibres). As mentioned earlier there are six short (ending in the lamina) and three long (ending in the medulla) R-fibres. The former are thus called short visual fibres (R(d) and R(s)), the latter long visual fibres (lvf). The six short R-fibres lie symmetrically in a circle in cross-sections at the level of the BM. The three lvf are in the center of the circle. The thick R(d)-fibres lie in the plane of symmetry, the two R(s)1 and the two R(s)2 being opposite to each other. The lvf are in a plane at right angles to the symmetry plane. This R-fibre pattern remains constant until the EPL is reached, where the R(s)-fibres end (Fig. 8).

In Golgi preparations three short visual fibre types can be distinguished (Fig. 9). The thick fibres (diameter 2–3 µm) are called deep retinula cells (R(d)) because they terminate in EPL(B) of the lamina. They have only small lateral protrusions but no real branches. The terminals are in the form of tassels with

Fig. 6a—d. Tangential sections at different levels of the retina and lamina. Left: micrographs; right: drawing; both from reduced silver impregnations. (a) Cross-section of retinulae just under the nuclei region of the visual cells. (b) Section through the proximal retina. Before the axon-bundles pass through the BM they show a regular orientation. The retinula-cell axons have quite a different pattern in cross sections. (c) Section through the CBL. After the torsion there is only a slight change in the axon-bundle-configuration inside the cartridges. (d) Section through EPL(B). Some of the R(s)-fibres are already missing. For a discussion of the diameters of the R-fibre-types see page 21

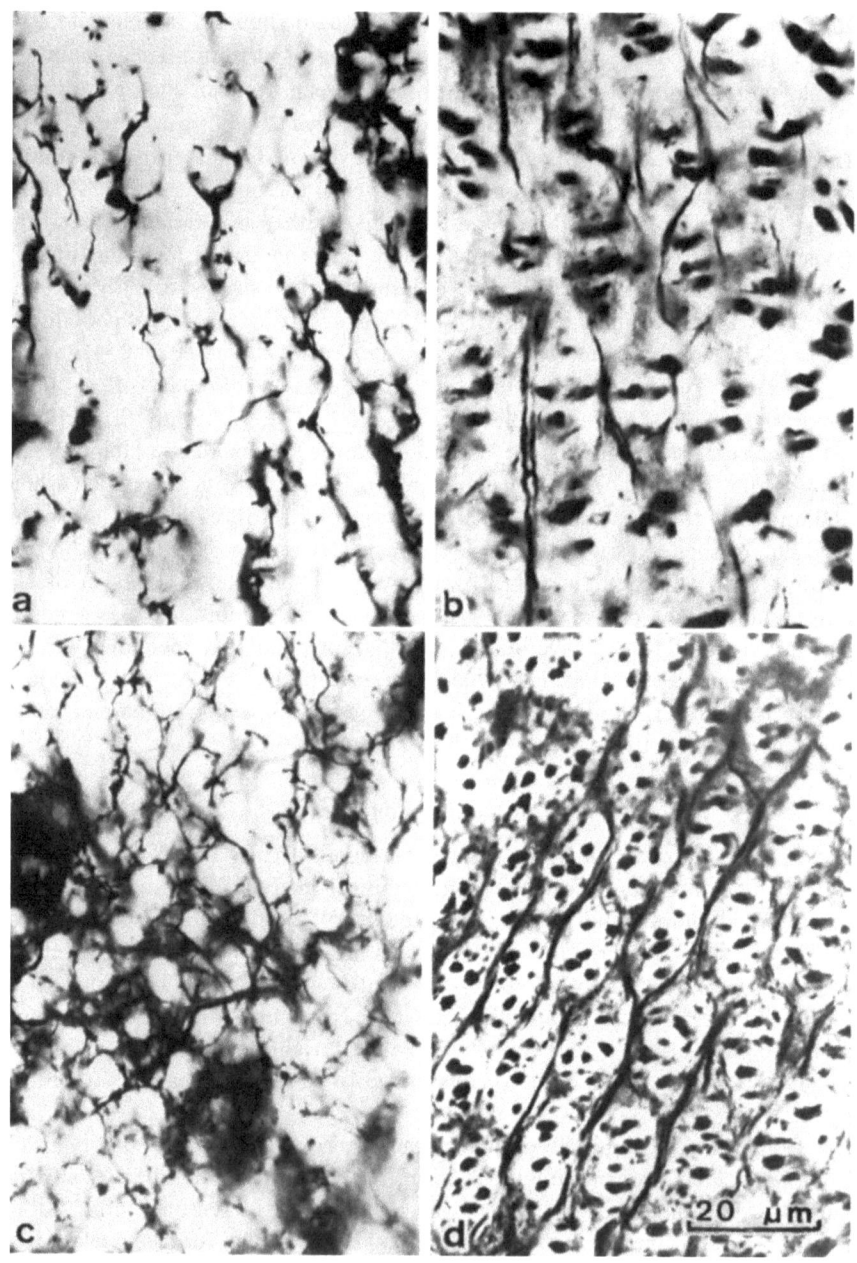

Fig. 7a—d. Tangential fibres at the level of the external plexiform layer (EPL) of the lamina resulting in an orientation of the cartridges along the x-, y- and z-axes. (a), (c) Selective silver impregnation (Golgi technique). (b), (d) Reduced silver impregnation

three to four arms. The four other R-axons (diameter 1–1.8 μm) are of two types but their terminals all lie in EPL(A) thus the name shallow retinula cells (R(s)). Those with forked terminals are called R(s)1, those with tapering ends R(s)2. Both

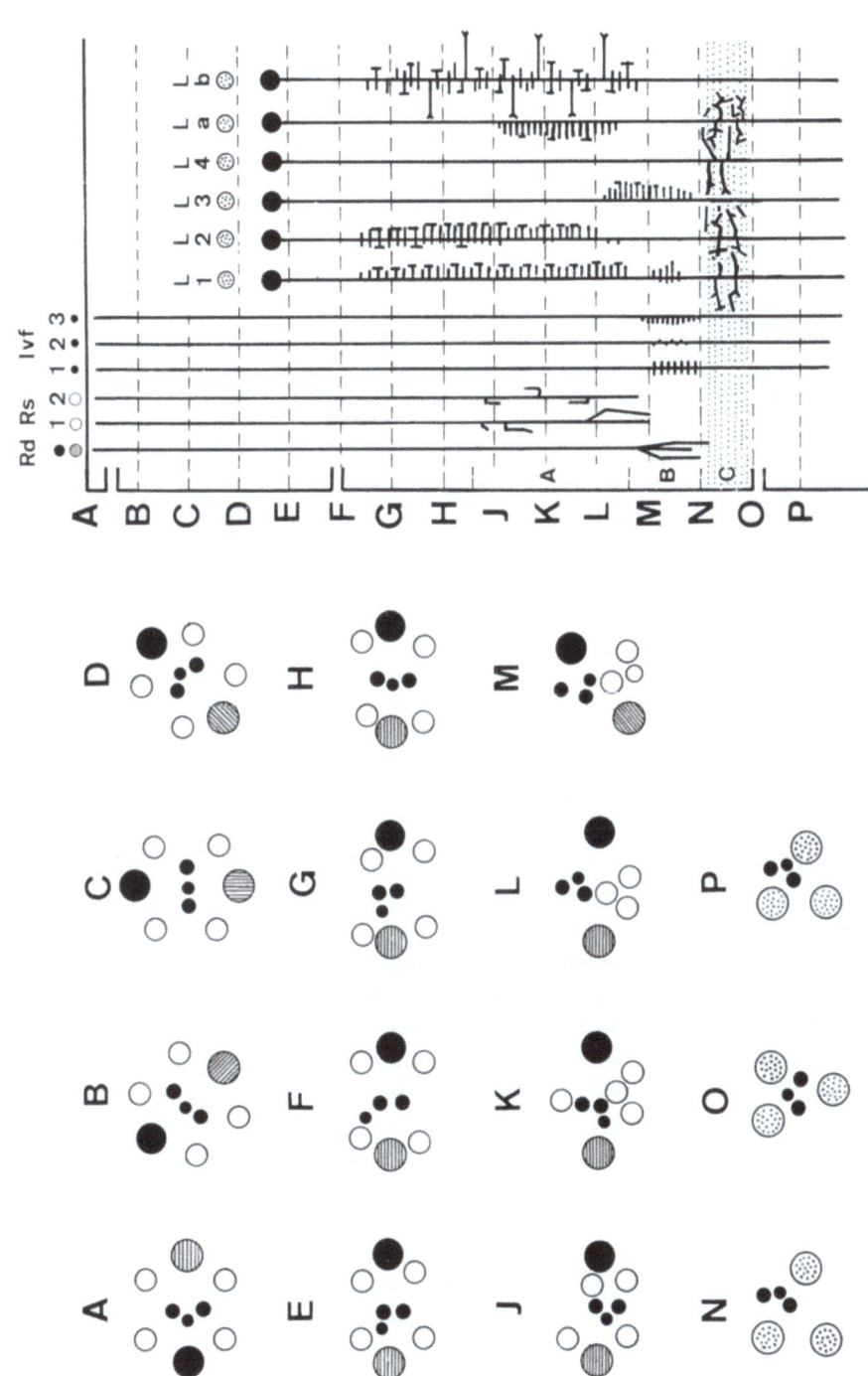

Fig. 8. Arrangement of fibres in the pseudocartridges and cartridges from the basal membrane (*BM*) to the beginning of the outer chiasma (*OCh*), from tangential section-series; reduced silver impregnation. The distance between each cross-section (each letter) is 10 μm. The bundles undergo a twist 180° clockwise or anti-clockwise between the basal membrane (BM) (cross-section *A*) and the cell body layer (CBL) (cross-section *E*). The L-fibres arriving into the cartridges are not drawn as their origin has not yet been determined at this level. $R(d)$ and $R(s)$ 1 and 2 reach two levels of the external plexiform layer (EPL): *L/M* and *M/N*. The long visual- (*lvf*) and *L*-fibres leave the lamina together in a six fibre-bundle

19

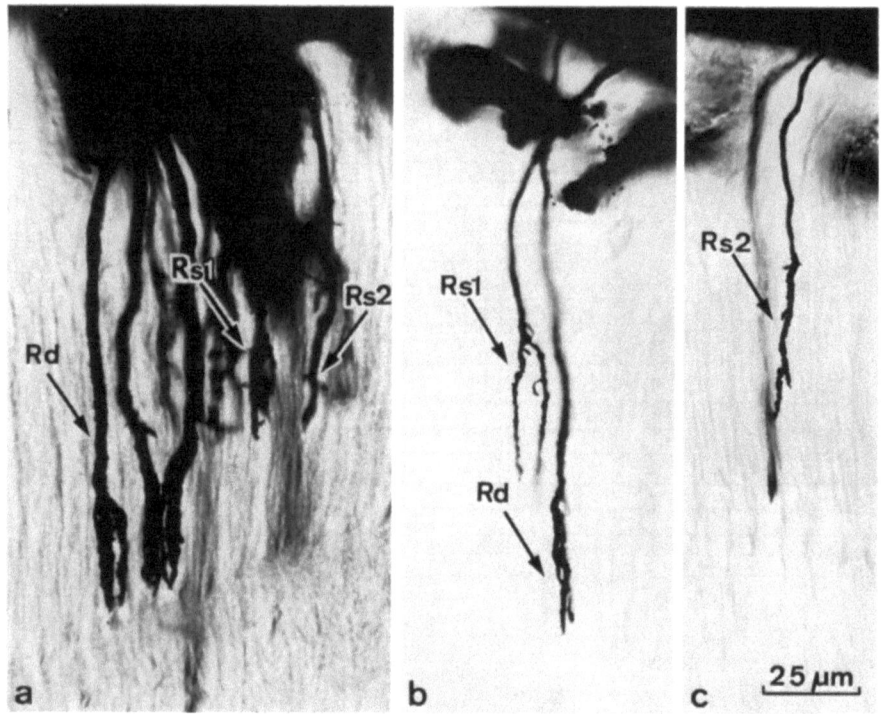

Fig. 9a—c. R-fibre endings in the lamina in selective silver impregnations (Golgi technique). (a) The three different R-fibre types according to their morphology, terminal depth and diameter. (b) $R(d)$ retinula-cell axon with deep, tassel-like ending, it also has the largest diameter. $R(s)1$ retinula-cell axon with shallow forked ending in EPL(A). (c) $R(s)2$ with unforked ending in EPL(A)

types have lateral branches in EPL(A). The light microscope techniques we used do not give information as to where there might be synaptic contacts with other fibres.

To summarize, a pseudocartridge is composed of six short visual fibres of three types, two fibres per type and of three long visual fibres. The long visual fibres have the smallest diameter (0.5–0.8 μm) and are usually of three types in the worker bee as well as in the drone. Irregularities are found mainly in marginal regions. In Golgi preparations all types have the same diameter which is quite constant from the BM to their terminals in the medulla. The types are distinguished by the depth of their terminals and the arrangement and shape of their spines (Fig. 11).

Lvf (1) is densely covered with spines in the EPL(B), the spines having a length of one fibre-diameter and being radially arranged. Lvf (2) has also spines in the EPL(B) but they measure only half the diameter in length. Lvf (3) has spines in the same layer (EPL(B)) but unilaterally only and averaging about twice the diameter in length. Lvf (1) contacts medulla neurons in the second medulla layer, its terminal is a knot-like swelling. Lvf (2) ends with small knobs in the first medulla layer. Lvf (3)'s diameter shows a sudden increase shortly after the beginning of the medulla and while it has tree-like branches in the second layer, the main branch leaves this tree-like structure, runs for about 25 μm tangentially and ends centrifugally in the first layer.

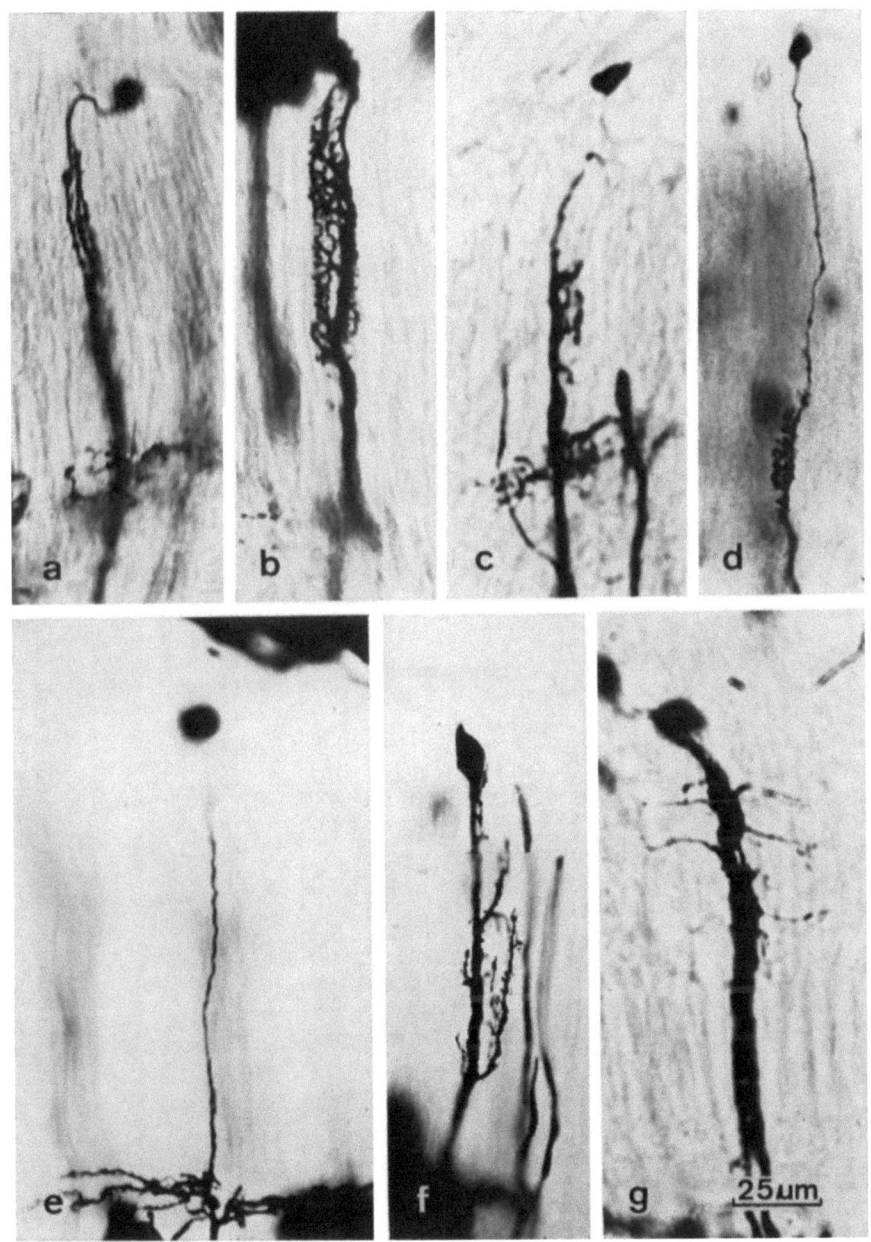

Fig. 10a—g. L-fibres of the lamina (selective silver impregnation; Golgi technique). (a) *L(1)*, (b) *L(2)*, (c) *L(a)*, (d) *L(3)*, (e) *L(4)*, (f) *L-fibre* with long forked branches, (g) *L(b)*. A detailed description of these fibres may be found in the text (pages 21 and 25)

Monopolar Cells (L-fibres). In Golgi preparations six types of L-fibres can be distinguished (Figs. 10, 11). As criterion the branching in the lamina, the shape and the depth of the terminals are used. Some of them strongly resemble *Musca* L-fibres. For this reason (and no other) those bee L-fibres thought similar to

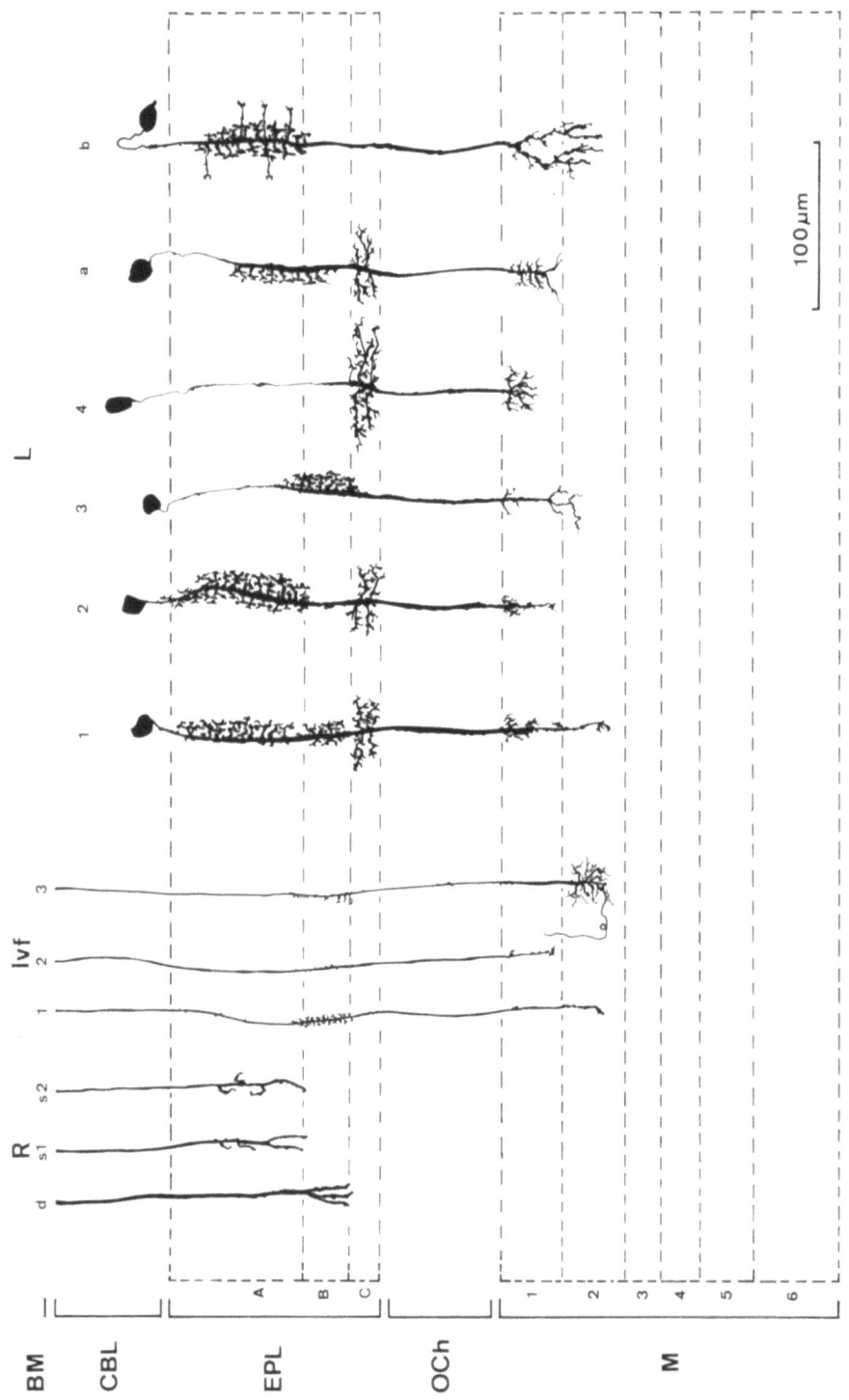

Fig. 11. R- and L-fibres as seen in Golgi preparations. $R(d)$ retinula-cell fibres with shallow ending in $EPL(A)$. Lvf (long visual fibres 1, 2, 3) run through the lamina and the outer chiasma. They end at two levels in the distal part of the medulla. $L(1)$, $L(2)$, $L(3)$, $L(4)$ are monopolar cells comparable to the L-neurons of dipterans. $L(a)$ and $L(b)$ are found in the bee only. For detailed description see text

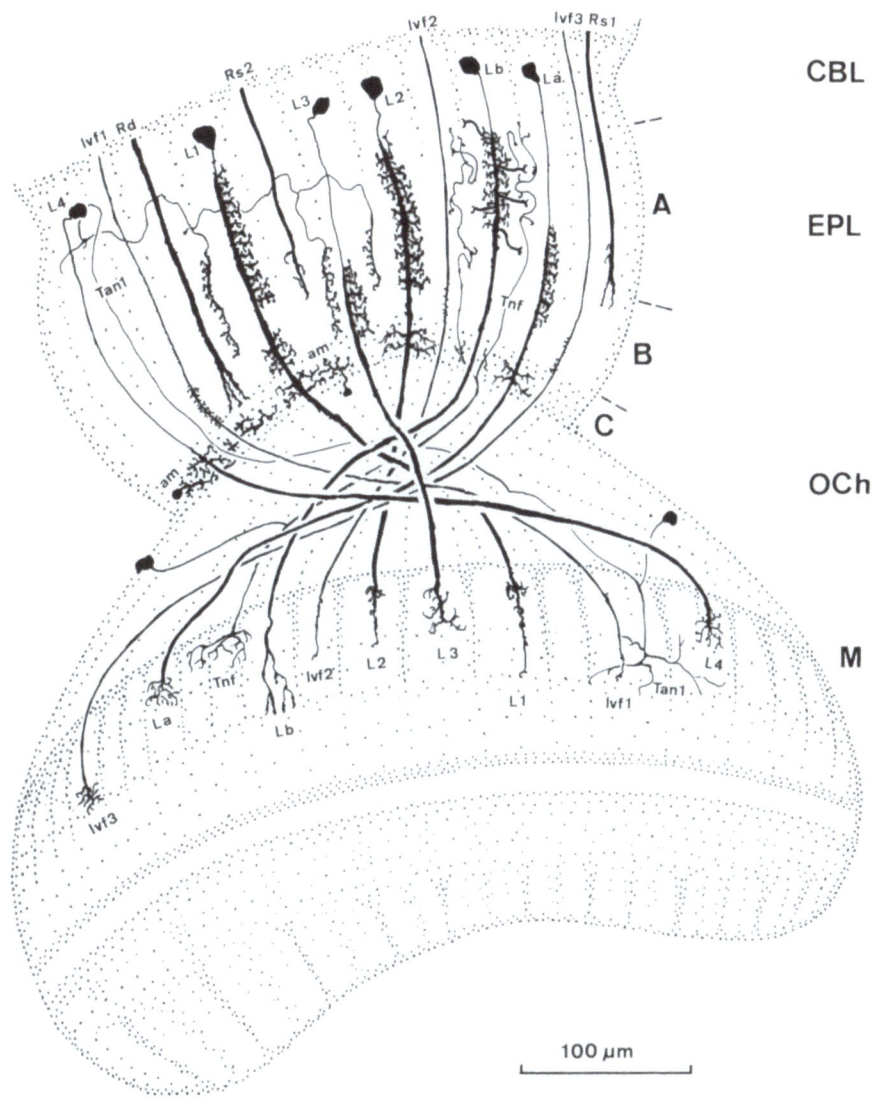

Fig. 12. Description of lamina and lamina-medulla elements from horizontal Golgi-prepared sections. $R(d)$, $R(s)1$ and $R(s)2$ end at two different levels in the lamina (*EPL*). The three long visual fibres (*lvf* 1, 2, 3) run through the lamina and the outer chiasma (*OCh*) without interruption. They end at two levels in the distal part of the medulla (*M*). The six morphologically different L-fibres $L(1)$—$L(4)$, $L(a)$ and $L(b)$ all have their cell bodies in the CBL being the distal part of the lamina. $L(1)$—$L(4)$ are very similar to the L-fibres of *Musca*. $L(a)$ and $L(3)$ differ only by the former's lateral branches in EPL(C). $L(b)$ has bilateral branches in $EPL(A)$ which can be compared to those of $L(1)$ and $L(2)$ and long collaterals with forked ends. The L-fibres end as do the lvf's on two levels in the peripheral medulla. The tangential and centrifugal elements have their cell bodies either in the cell body layer of the outer chiasma or further centrally. Some centrifugal components reach the lamina. The axon of $Tan(1)$ branches in the $EPL(A)$ and from these branches collaterals run parallel to the cartridges. They carry spines in $EPL(A)$, suggesting possible synaptic connections with the cartridges. $T(nf)$ splits in $EPL(C)$, the ends have only few branches in $EPL(A)$. The amacrines (*am*) are mainly found in $EPL(C)$, their axons run, branching often, through the fibre-dense $EPL(C)$

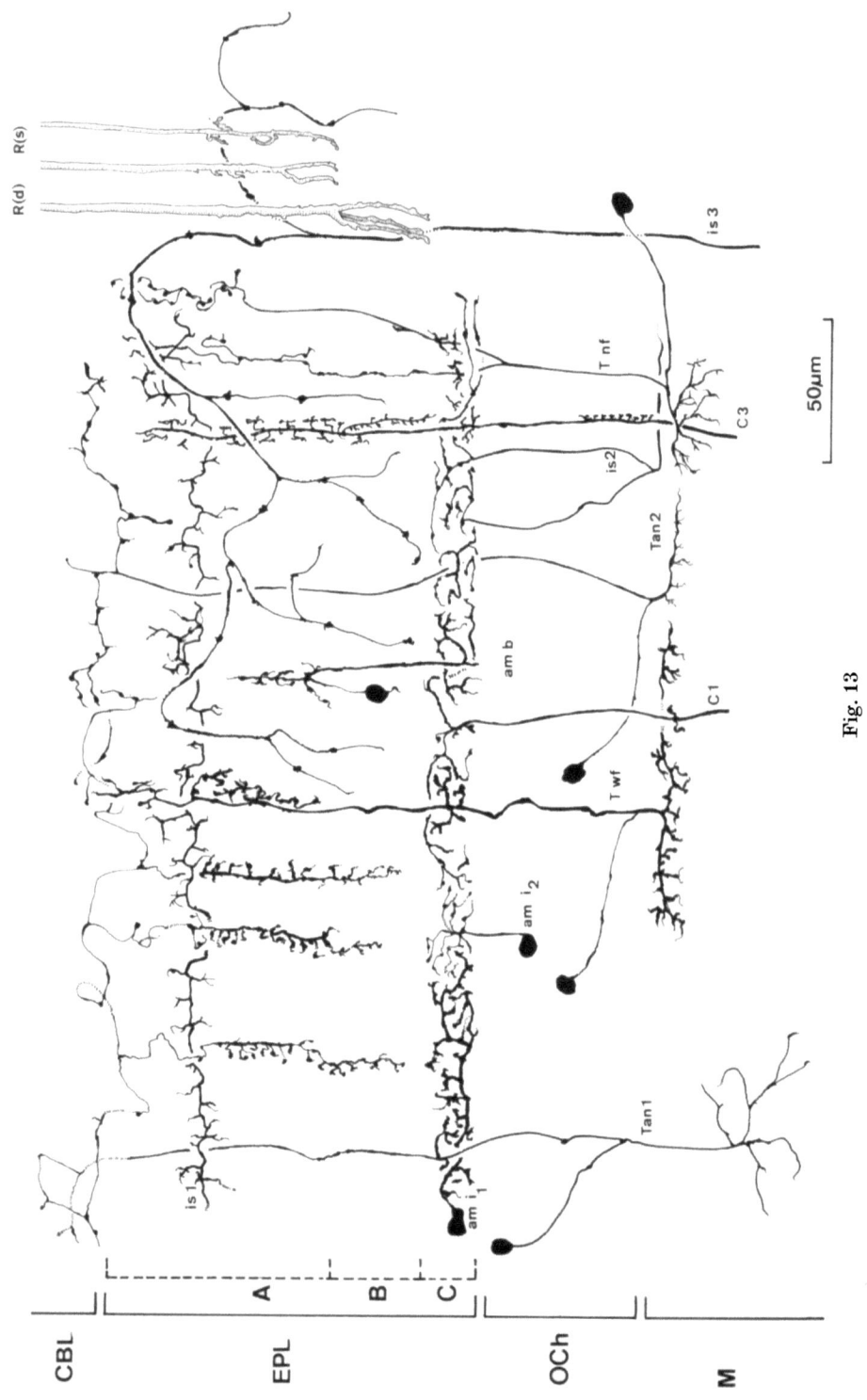

Fig. 13

Musca L-fibres were given the same symbols; the others have small letters as symbols.

The most outstanding feature of the L-fibres is the branching-pattern. There is radial, uni- and bilateral branching either at one or two levels or along the whole fibre. L(1), L(2) and L(b) have lateral branches for the whole length of (EPL(A)), possibly with synaptic connections to R-fibres. L(1) has additionally bilateral branching in EPL(B); L(1) and L(2) have lateral branching in EPL(C). L(b) can easily be distinguished in frontal sections by its partly 20 μm long collaterals occuring bilaterally in the EPL(A) next to short branches. Their terminals are forked. L(3), L(4) and L(a) can be called small monopolar cells as they have few or no lateral elements in EPL(A/B). L(a) alone has lateral elements in EPL(A). L(4) and L(a) have bilateral branching in EPL(C), L(3) unilateral branching in EPL(B). In EPL(C) we find long lateral branches (up to 80 μm) in horizontal sections belonging to L(4). Up to this point its central fiber has a diameter of 1 μm which then increases to several μm.

Tangential Fibres (T-fibres). Four types of tangential fibres are regularly found (Figs. 13 and 14). They all have cell bodies between the lamina and the medulla, most of them lying in the chiasma, near the medulla. Occasionally some are found in the cell body layer at the periphery of the chiasma.

There is a characteristic division of the main fibre into two branches at the level of the medulla surface. One branch ends in the distal medulla, the other in the lamina.

Tan(1) has its cell body either in the cell body layer next to the chiasma or at the surface of the medulla. It branches in the proximal third of the medulla; one arm has a few spine-less branches and ends centripetally in the second medulla layer (Fig. 15a). The second arm runs straight to the CBL, (Fig. 14a) then splits into many "garlands" occuring through the whole lamina parallel to the cartridges. These "garlands" have unilaterally small knotty protrusions which suggest synaptic connections with the cartridge-neurons.

Tan(2) has a similar shape to Tan(1) in the lamina, but without the "garlands" (Fig. 14g) moreover their horizontal extension in CBL and EPL is smaller. The medullary arm has short branches, running parallel to the surface and ends conically (Fig. 15c).

T(wf) cell bodies are grouped in the outer Chiasma, near the medulla surface. One arm innervates the distal medulla with a bilateral broad ending. The few branches carry short, thick protrusions (Fig. 15d). In the lamina portion of this

Fig. 13. Amacrines, centrifugal- and tangential cells from Golgi preparations. *am(b)* bistratified amacrine with branches at two levels. *am(i)* amacrine of inner stratum with branches in the *EPL(C)*. *C(1)* centrifugal type 1, terminal branches in the *EPL(C)*. *C(3)* centrifugal type 3 with short bilateral branches, the axis- fibre reaches the *CBL*. *Tan(1)* tangential cell 1 with garland-like branches parallel to the cartridges. *Tan(2)* tangential cell 2 with spine-covered branches in *EPL(A)*. *T(nf)* narrow-field tangential cell with short tree-like ending. *T(wf)* wide field tangential cell with extensive terminal branching in the medulla. *is(1)* incerta sedis, tangential of outer layer (lamina). *is(2)* incerta sedis, tangential of inner layer. *is(3)* incerta sedis with diffus branching. *R(d)* deep retinula cell axon; *R(s)* shallow retinula cell axon. The amacrines have their cell bodies at the periphery of the lamina at the level of *EPL(C)* or *(B)* and in the outer chiasma *(OCh)*. Centrifugal cells have their soma in the medulla *(M)* or in the cell body layer *(CBL)* at the level of the outer chiasma

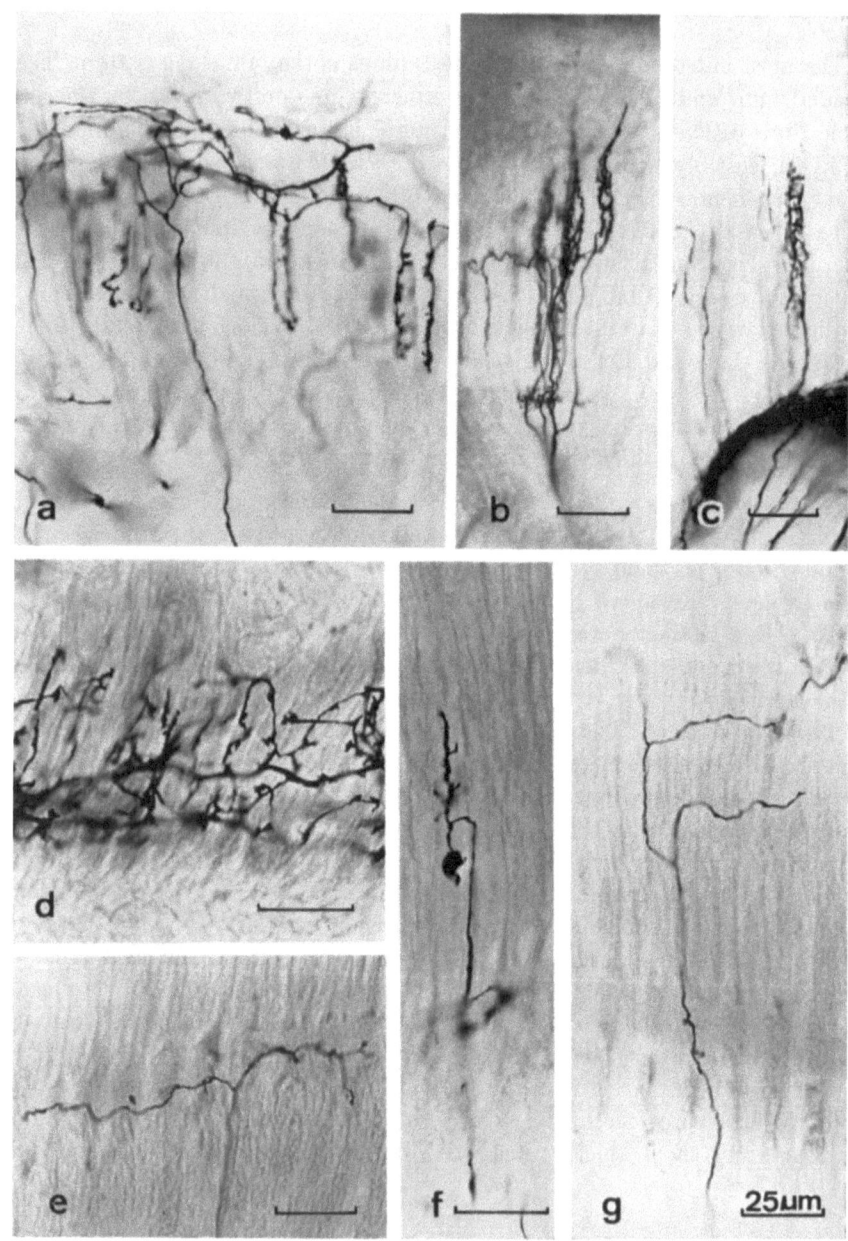

Fig. 14a—g. Amacrines, tangential- and centrifugal fibres of the lamina (selective silver impregnation; Golgi technique). See also Fig. 12 and 13). (a) Garlands of *Tan(1)*, (b) Elements of *T(nf)*, (c) Centrifugal fibre *C(3)* in the lamina, (d) Part of *is(1)* running through the whole lamina, (e) Branching of *C(1)* in EPL(C), (f) Amacrine *am(b)* with lateral branching at two levels, (g) *Tan(2)* with spineless lateral branching running through large parts of EPL(A)

neuron there are two levels of lateral branches (Fig. 13) which have club-like endings in EPL(C). The central fibre in the lamina also sends branches out at the level of the R(s) terminals. T(nf) cells originate from cell bodies in the outer

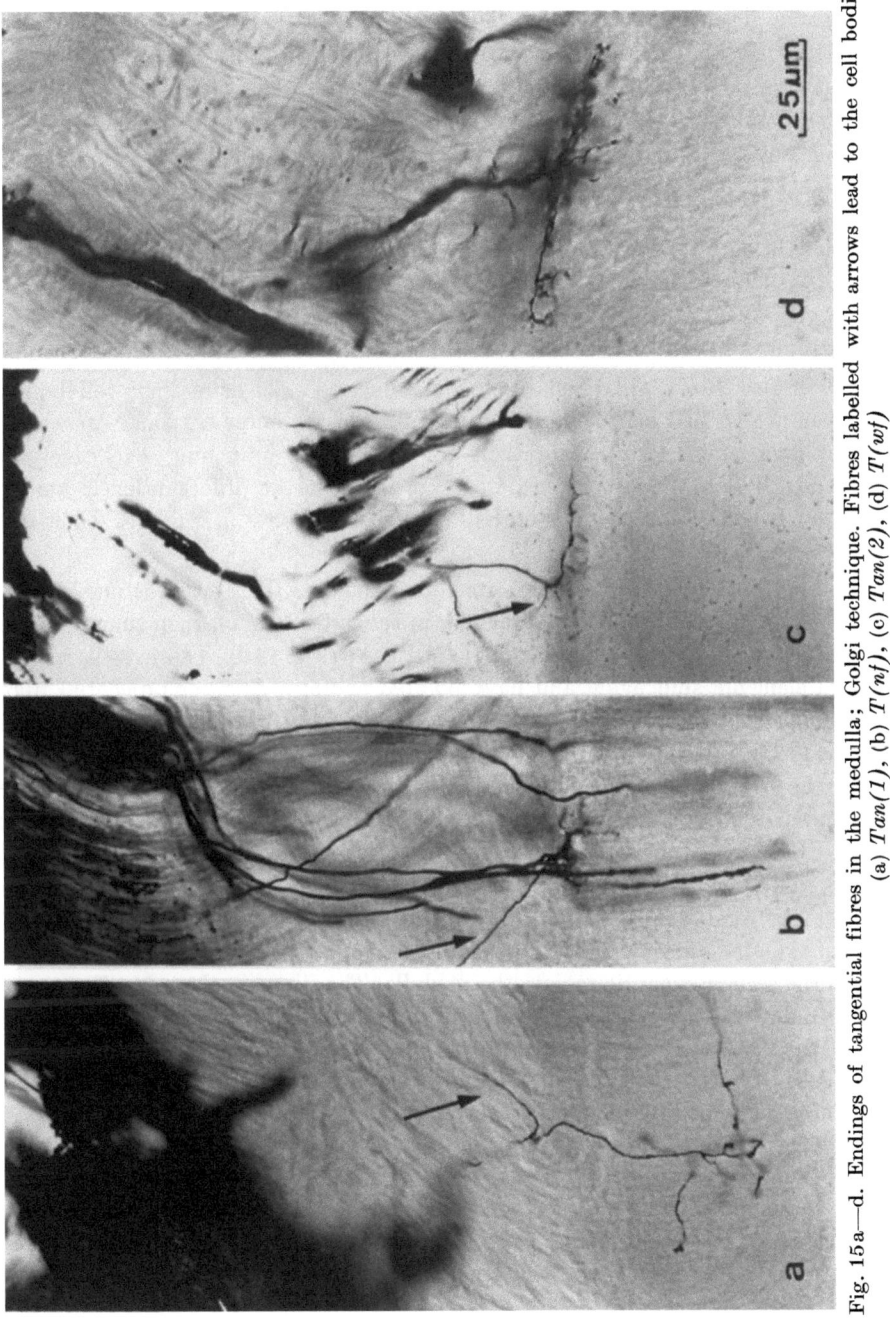

Fig. 15a—d. Endings of tangential fibres in the medulla; Golgi technique. Fibres labelled with arrows lead to the cell bodies. (a) *Tan(1)*, (b) *T(nf)*, (c) *Tan(2)*, (d) *T(wf)*

chiasma or in the neighbouring cell body layer. Their terminals at the medulla surface occupy only a small space (Fig. 15b). The lamina central fibre splits into two arms which then end at the distal part of the EPL with only a little branching (Fig. 14b). As in T(wf) there are short lateral protrusions in the EPL(C) and twisted elements in EPL(A).

Centrifugal Fibres (C-fibres). These originate either from cell bodies inside the medulla or from more centrally. The exact localisation of C(1) and C(3) cell bodies is not known. C(1) travels from the inner medulla through the outer chiasma into EPL(C) where it branches bilaterally over a distance of 150 μm (Figs. 13, 14e). Additionally there are other short straight elements at the medulla-surface, radially arranged.

The C(3) fibre also leaves the medulla in a centrifugal direction but reaches the CBL (Figs. 13, 14c). It has short lateral elements in the proximal outer chiasma, in EPL(C) and EPL(A/B). The former are unilaterally distributed short spines with T-shaped ends. Side elements in EPL(C) are radially arranged as are those in the EPL(A/B) with T-ends. One thin branch with minute spines leaves the main axis in the EPL(B) and runs centripetally to the EPL(C) where it ends.

Amacrines (am). All regularly found amacrine cells have their cell bodies between the EPL(B and C) and the distal part of the outer chiasma. Amacrines of the inner lamina, am(i), have their uni- or bilateral branchings exclusively in the EPL(C) (Fig. 13). Amacrines with side-elements at different levels am(b), have far reaching branches in the EPL(C) and short ones in EPL(A). Their cell bodies lie in EPL(B) (Figs. 13, 14f).

Incerta Sedis (is). These fibres are thus named as their origin is not known. Is(1) is a tangential-fibre type branching in the outer lamina, then running more or less straight through it to reach the EPL(A) (Fig. 14d). The irregular short lateral elements split again and have pointed endings. Another tangential fibre is is(2). It has lateral elements in the inner lamina and possibly originates in the cell body layer at the level of the outer chiasma. The main fibre runs from there along the medulla surface and divides in the EPL(C) into two arms. Is(3) is a diffuse tangential fibre coming from the medulla and branches bilaterally throughout the EPL. The tangential arms have few branches with regular but widely spaced knots.

5. Outer Chiasma

An axon bundle leaves each cartridge and passes through the outer chiasma to reach the second synaptic region, the medulla. The fibres in these bundles are of the lv-, L-, T- and C-types. Tangential sections of reduced silver impregnated tissue show 6–9 axons per bundle. Each bundle is separated from other by neuroglial cells having their bodies at the limit of lamina chiasma.

The lamina-medulla projection is similar to the retina-lamina projection dealt with earlier. The crossing in the horizontal plane is such that the cartridge pattern of the lamina is projected mirror-like onto the medulla (Fig. 12). Vertical sections show a torsion of the bundles around their own axes, and the torsion occurs in both directions. In the proximal part of the chiasma lie the cell bodies of tangential and transmedullar cells as well as of neuroglial cells.

6. Medulla

The medulla surface is covered by a dense tangential-fibre net forcing the incoming lamina fibres into a hexagonal pattern similar to the lamina pattern (Fig. 17). There are 6 layers in the medulla (Fig. 16). Lv- and L-fibres end in layers 1 and 2.

Terminals of Long Visual Fibres (lvf). The three lvf's of each retinula reach their terminations in the first two medulla-layers without branching (Fig. 11).

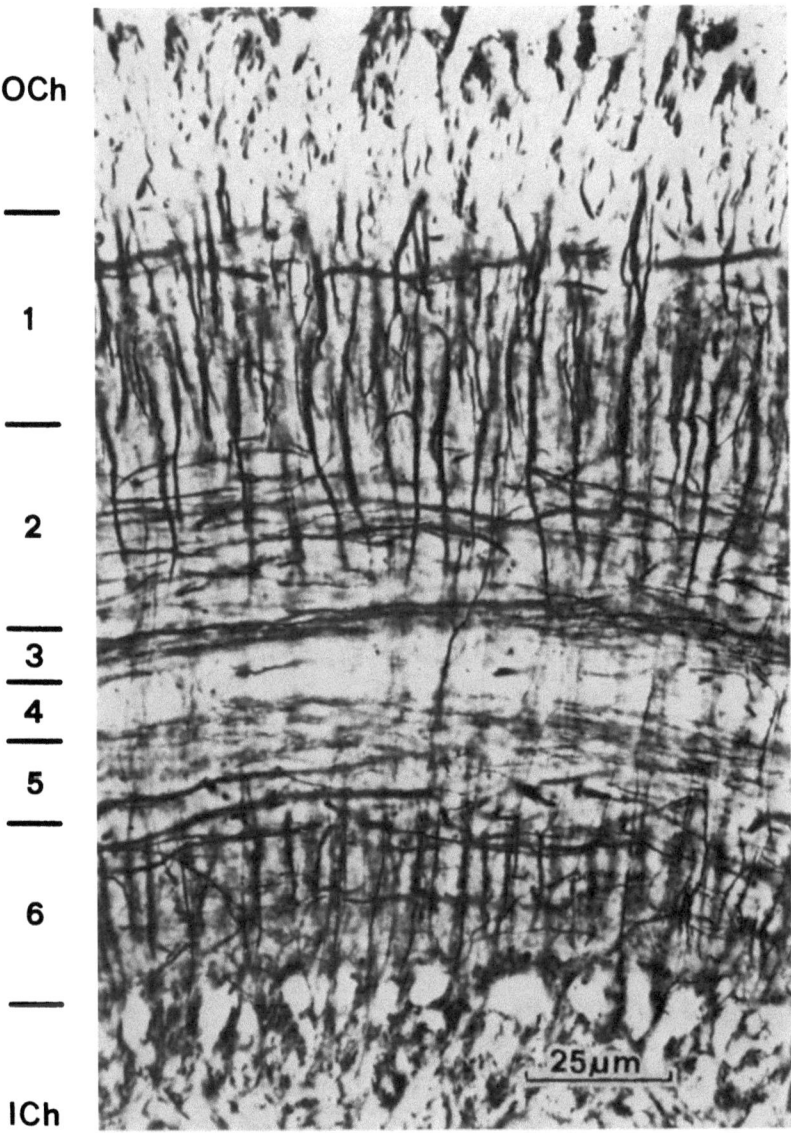

Fig. 16. The layers of the medulla, reduced silver impregnation, vertical section. *ICh* inner chiasma, *OCh* outer chiasma, 1–6 layers of medulla. Terminals of lv- and L-fibres in layer 1 and 2

Lvf(1) terminates in medulla layer 2 (Fig. 18a). Lvf(2) ends with small knobs in layer 1 (Fig. 18b). Lvf(3) increases its diameter on reaching the medulla and has tree-like branches in layer 2, but the main fibre continues tangentially for about 25 μm and finally ends centrifugally in layer 1 (Fig. 18b).

Monopolar-cell Terminal (L-fibres). The L-fibres all end in medulla layer 1 or 2 (Fig. 11). Upon entering the medulla L(1) sprouts radial branches, the length of which decrease with greater depth (Fig. 18c) until they disappear completely at the boundary between layers 1 and 2. The main fibre tapers out in layer 2. L(2) ending is similar to L(1)'s (Fig. 18d) though it is shorter, stockier and

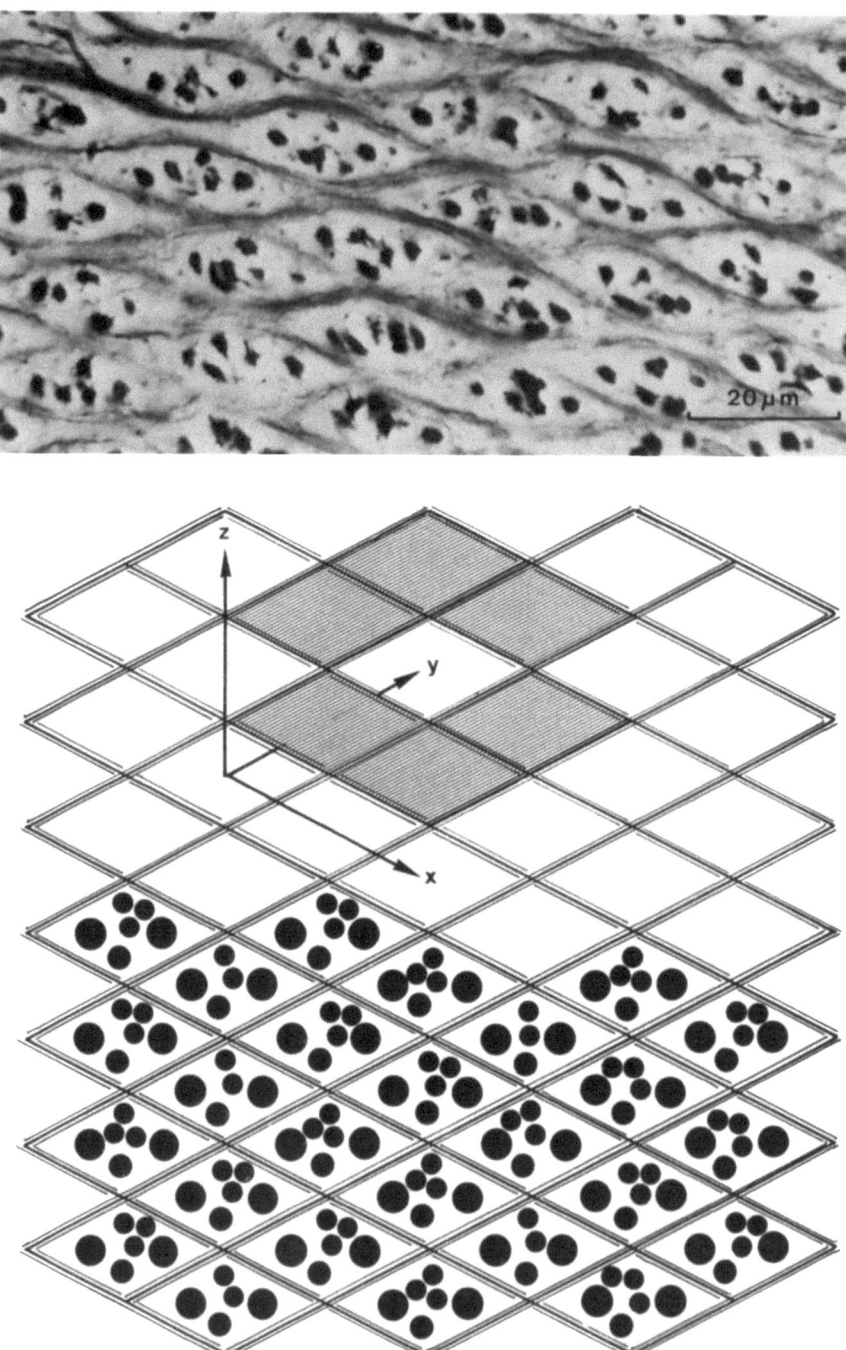

Fig. 17. Tangential section through the outer layer of the medulla. Photograph of a reduced silver impregnated section (above) and diagram (below). The visual cell and L-fibres reach the medulla through the outer chiasma in bundles of six. The glia- cell separated fibres show a clockwise or anti-clockwise twist at the level of the chiasma. The hexagonal pattern at the beginning of the medulla is similar to the lamina pattern and is brought about along the x-, y- and z-axes by tangential fibres. The lvf and L-fibres can be distinguished by their different diameters

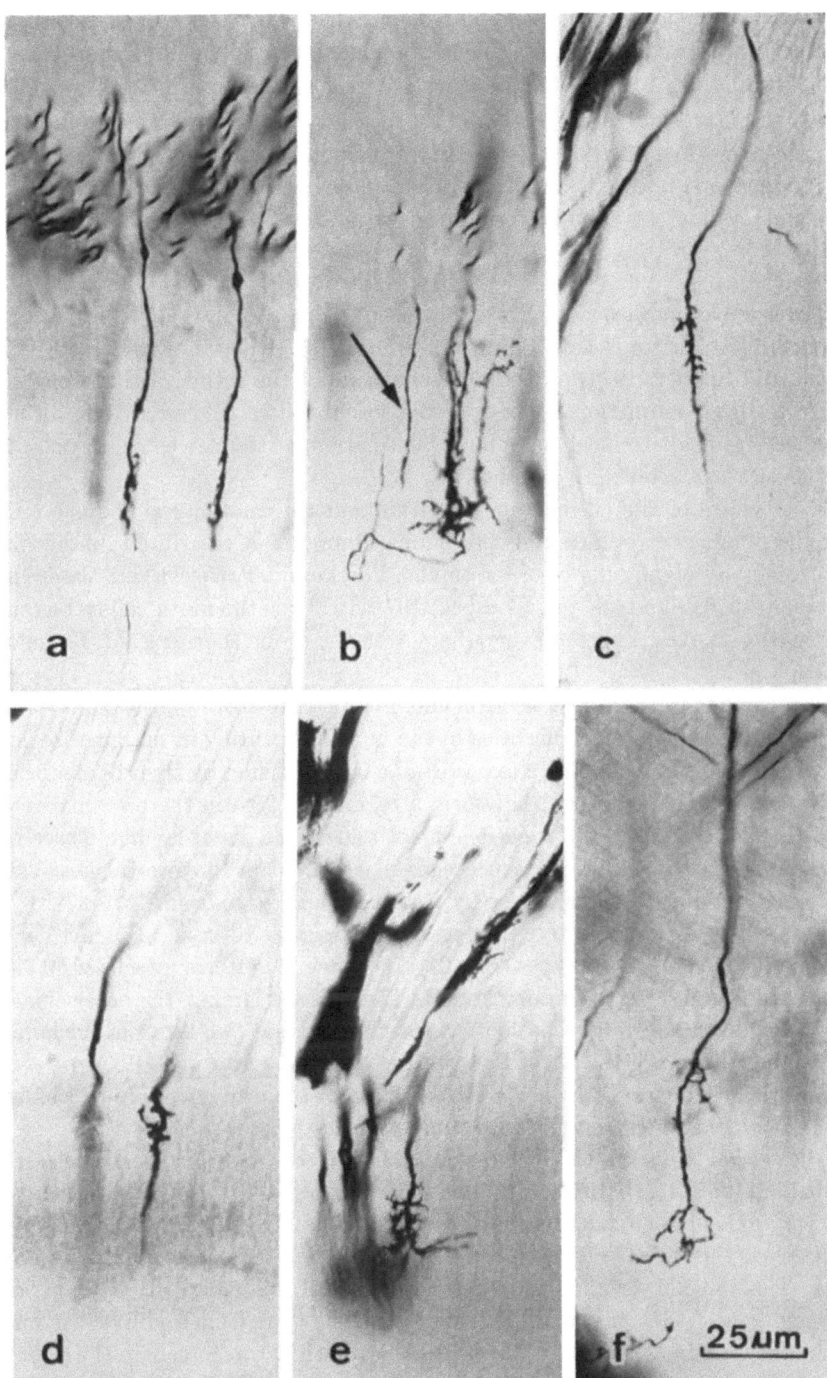

Fig. 18a—f. Long visual- and L-fibre terminals in the medulla, Golgi technique. (a) lvf(1) terminal. (b) lvf(2) terminal, marked by the arrow. Just next to it the branching of lvf(3) with the centrifugally-oriented main fibre running to the first layer, (c) The L(1) terminal lies in the second layer, (d) A L(2) terminal looking similar to the L(1) terminal, but ending in layer one, (e) End of a L(a) fibre, (f) End of a L(3) fibre with terminal branches reaching into layer two

sometimes has knobs at the end. It reaches only as far as layer 1. L(3)'s diameter is reduced after entering the medulla where it has short bilateral branches. The terminals fork into a few lateral elements which themselves have short smooth branches in layers 1 and 2, the branches may reach 20 µm (Fig. 18f). The endings of L(4) and L(a) reach the first layer of the medulla. L(4) shows branches radiating from one point (Fig. 11); the ending of L(a) looks similar to a bottle brush (Fig. 18e). L(b) divides in two arms each with only a few branches at the beginning of the medulla and terminates at the bottom of the second layer (Fig. 11).

IV. Discussion

The recently published electron microscopy papers (Varely and Porter, 1969; Perrelet, 1970; Skrzipek and Skrzipek, 1971, 1973; Gribakin, 1967, 1969, 1972; Menzel and Snyder, 1974) on the ultra-structure of the retina of the bee do not suffice to draw a functional model of the visual-cell connections. It is difficult to detect from anatomical data how information on the e-vector of polarized light or on the wave length is processed.

It is typical of the current uncertainty about the anatomy of the bee retina that the number of retinula cells per ommatidium is not clearly established: are there generally eight cells in one retinula (Varela and Porter, 1969; Varela and Wiitanen, 1970; Skrzipek and Skrzipek, 1971, 1973); is the ninth cell to be found only in certain eye-regions or everywhere as in the drone (Perrelet and Baumann, 1969b)?

According to the light and electron microscopy there are always 9 axons leaving each retinula, the only irregularities in the number occuring in marginal regions.

As for the visual cells, the 9 axons can be distinguished by their diameter and their position in the bundle. They form a regular pattern in the proximal retina (just above the BM), in the zona fenestrata and in the distal lamina. Three thin axons lie on a straight line; in a perpendicular plane run the two thickest axons which are surrounded on two sides by medium diameter-axons (Fig. 5).

From reduced and selective impregnations the nine retinula cell axons can be attributed to the following types (Fig. 21): the three thin fibres (diameter: 0.5–0,8 µm) are long visual fibres running straight through the lamina, the outer chiasma and ending in the medulla. The short visual fibres end at two levels in the lamina, two of them have a diameter of 2–3 µm, the remaining four of 1–1.8 µm.

The Golgi preparations allow a classification of the different fibre-types based on terminal-structures and end-branching-patterns.

The visual cells of the bee-retina can also be morphologically classified (Gribakin, 1967, 1969, 1972). The two type I cells have the most distal nuclei. The long axis of their microvilli are parallel. Type II cells (2 pairs, 250–300 µm long each as the type I cells) have their microvilli oriented perpendicularly to each other. Their nuclei lie about mid-retina. Type III cells are only 150 µm long, the long axes of their microvilli is lying 90° to the type I microvilli. Their nuclei are in the proximal part of the retina. The ninth retinula cell is the shortest (50–80 µm) and its nucleus is found just distal to the BM.

Intracellular electrophysiology has shown that the visual cells of the worker bee-retina are of at least three spectral types: the maxima of their sensitivity curves lie at 360 nm, 420 nm and 540 nm (Autrum and Zwehl, 1964; Goldsmith, 1964) but we have no solid evidence as yet indicating how these types correspond to the anatomical retinula cell-types.

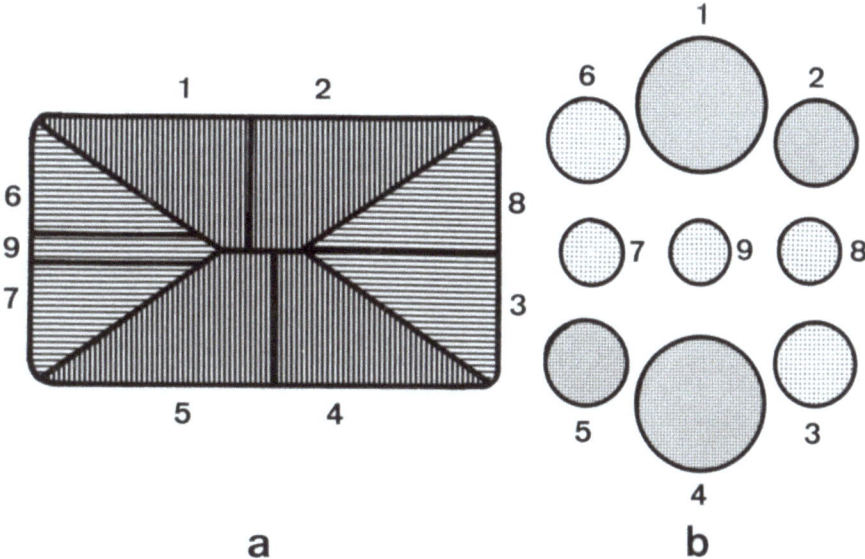

Fig. 19a and b. Possible correspondence between visual cells and axons. (a) Proximal cross-section of a rhabdom. Two facing rhabdomeres always have the same microvilli-direction with the exception of the ninth cell. The perpendicularly oriented microvilli are marked with different signatures. Visual cells with equally oriented microvilli are: 7/8; 6/3; 2/5; 1/4; 9/—. (b) Projection of the microvilli- pattern onto the visual-cell axons. According to this model two morphologically identical cells send the same sensory information to the nervous system

Selective adaptation methods based on the movement of the retinula-cell screening pigments and changes in the ER have not succeeded to this date.

Gribakin (1969, 1972) reports that after strong illumination with monochromatic light he finds such ultra-structural differences in the microvilli as to be able to identify morphologically three visual cell types: two maximally UV-sensitive cells, two with maximum sensitivity in the blue and four with maximum sensitivity in the green. Moreover Menzel and Snyder (1974) found after intracellular recording that the ninth cell must be a purely UV-sensitive receptor and that Gribakin's UV-cells must have a secondary sensitivity maximum in the green.

The neurohistological evidence reported in this paper also suggest three cell types plus the ninth cell. This conclusion is based on observations of the termination of retinula axons and of their diameter and positions within the pseudocartridge. Two retinula-cell axons (R(d)) have three to four-armed terminals in the EPL(B). Four axons (R(s)), two each having the same terminals, reach EPL(A). The remaining three axons pass through the lamina and end in the first two medulla-layers.

This classification is supported by the fact the cross sections of the axons of one bundle at the level of the BM and further proximally yields three groups with characteristic diameters. Electron microscopic investigations done by our group (Rothenbach, in prep.; Sommer and Wehner, in prep.) allow a reconstruction of the basal retinula and the above mentioned axon- bundle cross-section. Tangential sections show two large cell profiles in the proximal quarter of the retina, numbers 1 and 4 (Gribakin's, 1969, 1972 Type I). These cells are surrounded on both sides

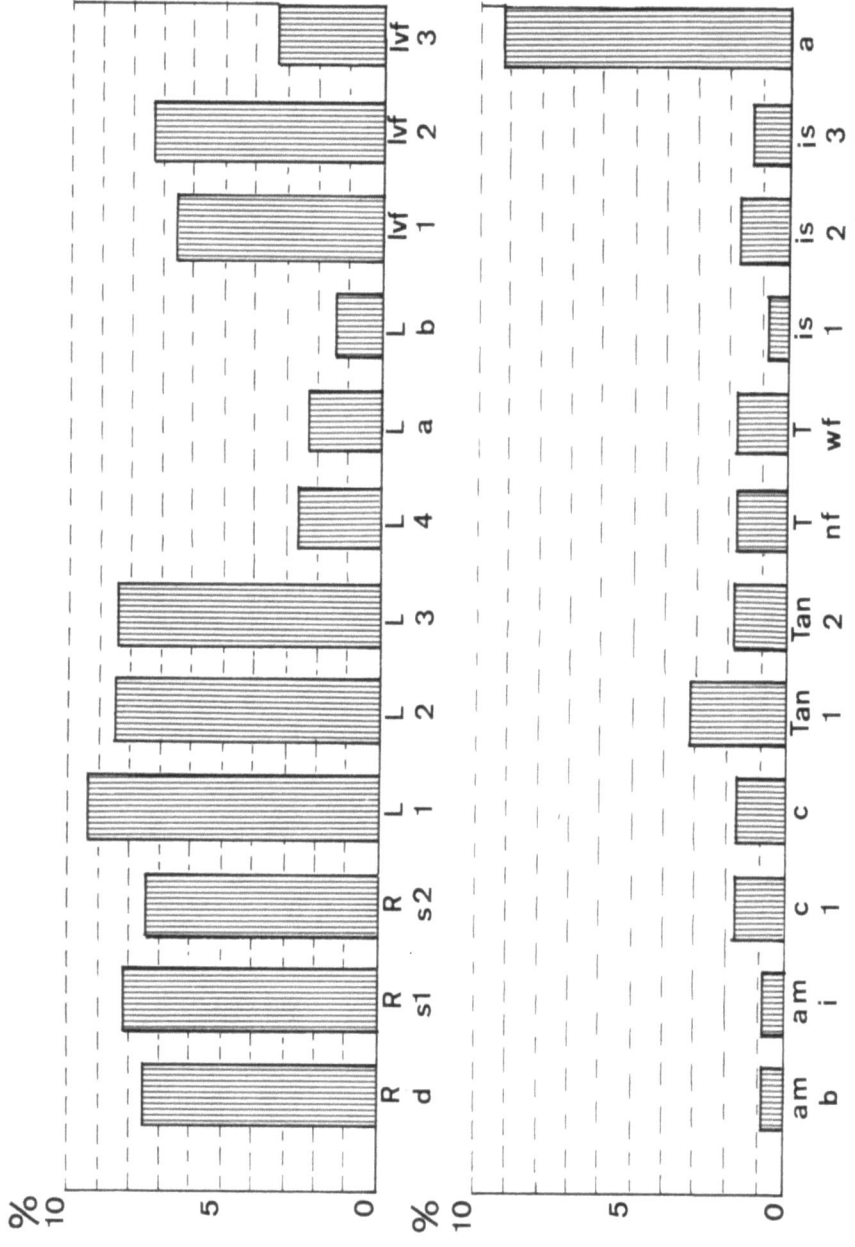

Fig. 20. Frequency of fibre-types in the lamina, from Golgi preparations. Based on 600 section series (10000 sections). The bars indicate in percent the frequency of the types. The sum of all types equals 100%. (see Table 1)

by four smaller cells (numbers: 2, 3, 5, 6, type II). At this level cells 7 and 8 (type III) are already axons. The space of cell body 7 is taken up by the ninth cell (Gribakin, 1972; Grundler, 1972), cell 7 has the most distal axon.

Shortly above the BM there is a retinula-cell pattern that can be directly projected on the axon-pattern of one bundle: two thick, four intermediate and three thin axons (Fig. 5). These results show clearly that retinula-cells with large cross-sections in the proximal retina (1 and 4) have correspondingly large axons.

34

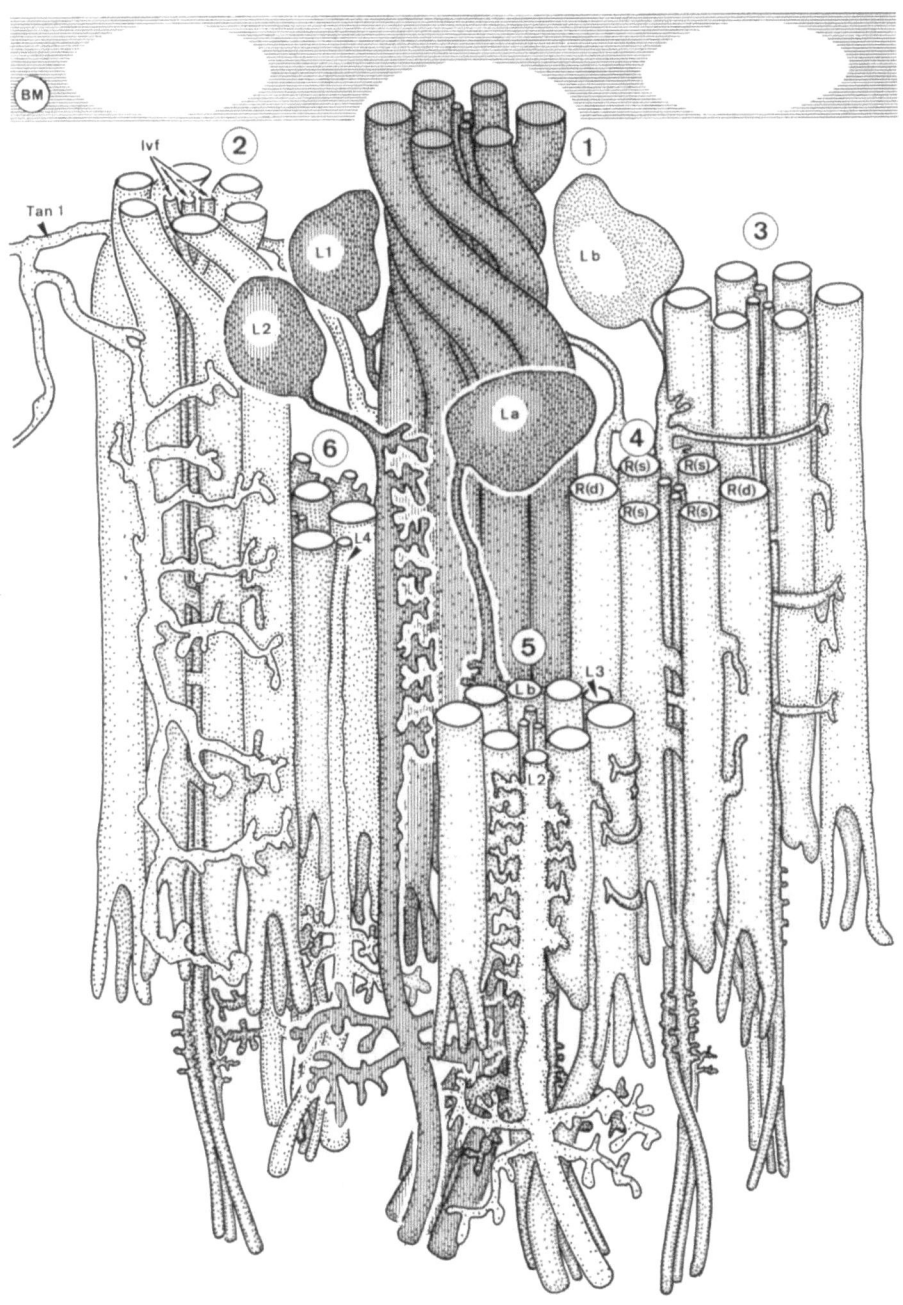

Fig. 21. Model of the components of cartridges in the lamina. Reconstructed from tangential section-series stained by reduced silver impregnation and the Golgi technique. *BM*, basal membrane through which the axon bundles (pseudocartridges) pass in a regular pattern. Bundle ① shows an anti-clockwise twist of 180° after passing through the BM. Six thick axons (short visual fibres) surround the three long visual fibres *(lvf)*. The *L*-fibres and *R*-fibres form together an optical cartridge. The L-fibres adhere closely to the R-fibres through branches which conceivably make synaptic contacts. The R-fibres $R(d)$, $R(s)1$ and $R(s)2$ end on two levels of the external plexiform layer (EPL) A and B. *L(2)* clearly has branches in the EPL(A) and a thicker fibre cross-section. Larger lateral protrusions lie in the EPL(C).

It is thus possible to correlate the e-vector sensitivities of the single visual cells to their axons in the lamina and medulla although a similar correlation of spectral sensitivities is not definitely possible at this stage.

If we accept Gribakin's results (1969, 1972), although they could not be reproduced by Grundler (1973) and have been criticized by Eguchi and Waterman (1973), we would have to call the cells of type II (2, 3, 5 and 6) green receptors. These are the four cells which have the same intermediate axon cross-section and have their terminals in the same lamina layer. Moreover it is further possible that the two type II cells with a common microvilli-direction might belong to one terminal-type (R(s)1 or R(s)2). If this were so, the other two cells with a microvilli direction perpendicular to these would then belong to the other type, R(s)2 or R(s)1 (Figs. 11, 21).

Cells 1 and 4 (type I) would have to be called UV-recpetors corresponding to R(d)-fibres. The ninth cell, according to Menzel and Snyder (1974) exclusively a UV-receptor, with microvilli perpendicular to type I cells (1 and 4) would leave the retina as one of the three thin axons and end in the medulla.

The two remaining cells (7 and 8, type III) would be blue receptors and have long thin axons (lvf) also ending in the medulla. Considering all known anatomical data we prefer this interpretation to the one which at first glance is more obvious, i.e. where the axon-types R(d), R(s) and lvf would belong to three spectral types of retinula-cells.

The retinula-cell axons of each ommatidium pass the BM and meet the L-fibres after clockwise or anticlockwise torsion to form the cartridges. Unlike the regular torsion in *Musca*—clockwise in the upper eye-half and anticlockwise in the lower (Braitenberg, 1966)—the direction of torsion in the bee appears irregular.

Reduced silver impregnations show that any one hexagonally-arranged ommatidium lies above an identically oriented hexagonally arranged cartridge. Thus there is a direct projection of the ommatidium into the cartridges. The same applies to the pseudocartridges (Fig. 8).

Three methods were used to determine how many L-fibres (monopolar cells, second order neurons) belong to one cartridge; methylene blue staining, reduced silver- and selective silver impregnation (Golgi technique). In the methylene blue

L(3) thickens only in the EPL(B) after sending lateral branches to the R-fibres; it has no lateral elements in the EPL(C). *L(1)* is hidden by the cartridge. In all, six fibres leave the lamina as a bundle, three L- and three lv-fibres. Bundle ② shows a cross-section after a torsion of only 90°. A portion of a tangential fibre shown *Tan(1)* has its cell body in the outer chiasma and divides into two arms, one innervating nearly the whole lamina at the distal level of the EPL by sending fine branches parallel and adhering to the cartridges. In cartridge ③ *L(b)* is illustrated which has short spines and longer lateral branches in the EPL(A). There are lateral elements from *R(s)* fibres too. *R(d)* has never been found to have lateral branches except small nipples at the proximal end, they usually end with three or rarely four arms. The *R(s)*-fibres both end in the EPL(A), forked or simple. Cartridge ④. The lvf's run straight through the lamina and end with the L-fibres in the two first medulla-layers. They have no lateral projections except short ones in EPL(C). The bilaterally arranged spines of *L(2)* are shown in cartridge ⑤. *L(a)* and *L(b)* are masked by the R-fibres with the exception of the lateral arms of L(b) next to one of the R(d)'s. Near cartridge ⑥ L(4) can be seen with its long thin axon reaching the EPL(C) and the typical long and branching lateral arms

Table 1. Frequency of the nerve-cell types of the lamina from Golgi preparations

	R(d)	R(s)1	R(s)2	lvf(1)	lvf(2)	lvf(3)
a	7.6	8.1	7.6	6.7	7.2	3.3
b	32.6	34.6	32.6	38.8	41.6	19.4

	L(1)	L(2)	L(3)	L(4)	L(a)	L(b)
a	9.5	8.6	8.6	2.7	2.4	1.4
b	28.5	25.7	25.7	8.6	7.1	4.3

	am(b)	am(1)	C(1)	C(3)	is(1)	is(2)	is(3)
a	0.9	0.9	1.8	1.8	0.9	1.8	1.4

	Tan(1)	Tan(2)	T(nf)	T(wf)	fibres of undetermined origin
a	3.2	1.8	1.8	1.8	8.2

a = Percentage share of one fibre-type compared to all fibre types.
b = Percentage share within a single fibre group.

series the monopolar cell bodies were stained and their number compared with the total number of ommatidia computed from the number of corneal lenses per eye. A mean value of 5400 ommatidia was determined from ten eyes, whereas the mean number of monopolar cells amounts to three times this value. As the monopolar cell bodies can definitely be distinguished from the nuclei of glia cells and of tracheoblasts it can be deduced that each cartridge must have three monopolar cells.

The same conclusions are reached by interpreting the results from tangential sections (reduced silver impregnation): there are six fibres leaving each cartridge, three of these have a diameter of 0.5–0.8 µm and are long visual fibres; the three others have a variable diameter (2–4 µm) and must be monopolar cells. The axons leaving the lamina run in tight bundles through the chiasma and are observed again to project in a mirror-like fashion onto the medulla surface in a hexagonal pattern associated with the dense tangential fibre net (Fig. 17).

The Golgi preparations allow the distinction of six L-fibres that are in part similar to the L-fibres of Dipterans.

As for the monopolar cell bodies, a count of cell fibre types can give some information as to which types form a cartridge. In roughly 600 Golgi preparations (or 10000 sections) all fibres were assigned to known categories and counted (Fig. 20, Table 1). The results show that the three visual cell fibre-types are equally frequent, and that the same is true for the L-fibres L(1), L(2) and L(3). The total number of L(4), L(a) and L(b) fibres are each only another of the total number of cartridges (Fig. 20, Table 1). This means that only every fourth cartridge would be found to have one of these. Lvf(1) and lvf(2) are equally frequent in the cartridges, whereas lvf(3) appears in only half this number. One reason for this

Table 2. Site of lateral elements of the different fibre types in the EPL of the lamina

	R(d)	R(s)1	R(s)2	lvf(1)	lvf(2)	lvf(3)
EPL(A)		×	×			
EPL(B)	×			×	×	×
EPL(C)						

	L(1)	L(2)	L(3)	L(4)	L(a)	L(b)
EPL(A)	×	×			×	×
EPL(B)	×		×			
EPL(C)	×	×		×	×	

	am(b)	am(i)	C(1)	C(3)	is(1)	is(2)	is(3)
EPL(A)	×			×	×		×
EPL(B)				×			×
EPL(C)	×	×	×	×		×	

	Tan(1)	Tan(2)	T(nf)	T(wf)
EPL(A)	×	×	×	×
EPL(B)	×			
EPL(C)	×		×	×

low lvf(3) count could be that the impregnation does not work as well with this thin fibre (diameter 0.5 µm) as with thicker ones. Amacrines, centrifugal- and tangential fibres should each amount to a fourth of the cartridge-number. Nine percent of the fibres found cannot be attributed to any known category.

In these interpretations of the results, one must be aware of the fact that a different affinity of the fibre-types to silver might influence the results. But usually the observed Golgi-based relationships are confirmed by other staining techniques.

The branching pattern of the different fibre-types in the external plexiform layer of the lamina suggest synaptic connections in light-microscopical sections (Table 2).

The three short visual-cell axon types innervate the EPL of the lamina at two different levels: R(d) fibres contact both EPL(A) and EPL(B), while the shorter lateral fibres of R(s)1 and R(s)2 only the EPL(A). All these short visual fibres are connected to L-fibres which themselves have branches at different levels of the EPL: L(1) in the EPL (A, B and C), L(2) and L(a) in the EPL(A) and C), L(b) in EPL(A) only, L(3) in EPL(B) and L(4) in EPL(C). The lateral elements of the L-fibres in the EPL(C) have a maximal extent of 80 µm, suggesting the possibility that they might contact as many as five neighbouring cartridges. Long visual fibres have spines of varying length at the level of the R(d)-terminals or the EPL(B).

Tan(1), the fibre having garland-like collaterals over large parts of the lamina running parallel to the cartridges, has spines at the level of the terminals of R(d) and R(s)-fibres only. All other fibres that seem to belong to more than one cartridge do not show such strict regularity as to the location of spines. One feature however, is common to all fibre groups. Each type belongs to definite layers of the lamina neuropil. At this stage we can only speculate as to the function of these tangential and centrifugal elements of the bee lamina.

Summary

The lamina and the directly connected parts of the visual system of the bee (*Apis mellifera*) have been investigated with the light microscope.

1. General Anatomical Features

Retina. Each compound eye is made of approximately 5400 ommatidia. Each ommatidium consists of 9 retinula cells. These can be morphologically attributed to four groups, the criterion being the position of their nuclei. A short distance above the basal membrane the retinula cells change to axons which project to the lamina.

Lamina. The lamina ganglionaris is the outermost of the three optic ganglia. It can be roughly divided in a cell body layer (CBL) and fibre layer (EPL) (neuropil). The CBL has a external trachea-rich region and a region containing the monopolar cell bodies (MCBL). The retinula short visual fibres (svf) terminate at different depths in the proximal part of the lamina. These terminals and a dense horizontal net of fibres divide the EPL in three sectors (A, B and C).

Outer Chiasma. Neuroglial cells separate the axon bundles from the lamina throughout the chiasma. The crossing of the fibrebundles in the horizontal plane is such that the pattern of the cartridges is projected into the medulla in a mirror-like fashion.

Medulla. The surface of the medulla is covered by a dense net of tangential fibres. The arriving lamina fibres make their way through these in a hexagonal pattern similar to the lamina-pattern. Selective and reduced silver impregnations reveal 6 different layers. Lv- and L-fibres end in the 1st and 2nd distal layers.

2. Retinula-(svf and lvf) and Lamina-fibres (L-, am-, T-, C-)

Short Visual Fibres (svf). In Golgi preparations three svf-types can be distinguished. Both thick fibres (diam 2–3 µm) are named deep retinula cells (R(d)) because of the location of their terminals in the lamina (EPL, B).

Except for small protrusions there are no collaterals. The endings consist of three to four thick short tassel-like branches. The four other R-axons (diam. 1–1.8 µm) have their terminals in the EPL(A), hence the name shallow retinula cells. There are two types, R(s)1 which has forked endings and R(s)2 with tapering endings. Both R(s) types have collaterals in the EPL(A).

Long Visual Fibres (lvf). The three long visual fibres are easily recognised by the small diameters (0.5–0.8 µm) which are constant from the retina to the medulla. They can be distinguished from each other by differing terminal depths, (lvf 1,2,3), the shape of their terminals and the disposition of their spines in the lamina.

The three types all have spines in the EPL(B) and EPL(C) but with different densities and lengths. The lvf's end in the two outer layers of the medulla, lvf 2 in the most distal one, lvf 1 in the next deeper one. Lvf 1 and 2 have a knot-like ending, lvf 3 ends in the same layer as lvf 1 and has a tree-like ending. A collateral of lvf 3 runs tangentially in the 2nd medulla layer for about 25 µm then ends centrifugally in the most distal layer.

Monopolar Cell Fibres (L-fibres). There are 6 distinct L-fibre types in Golgi preparations, the most striking differences being the branching patterns in the lamina and the terminals in the 2 distal medulla layers. L(1), L(2), L(a) and L(b) have collaterals in EPL(A) (the layer where svf R(s)1 and R(s)2 end). L(1) has additional collaterals in EPL(B), the layer where svf R(d) end. L(4) has extensive bilateral branching in EPL(C). In the same layer one finds L(1), L(2) and L(a) with bilateral branching. L(b) has in contrast to other L-fibres long (up to 20 µm) bilateral collaterals and short forked processes in EPL(A). Of all L-types only L(1), L(3) and L(b) reach the 2nd medulla layer. Each L-fibre can be distinguished from the others by the characteristic structure of its ending.

Amacrines, Tangential- and Centrifugal Fibres. The cell bodies of the amacrines lie between EPL(B) and (C) and the distal part of the outer chiasma. They all have branches in the EPL(B) and (C) confined to one or two horizontal planes.

The four frequently found tangential fibres have their cell bodies between lamina and medulla, mostly in the portion of the outer chiasma next to the lamina. The central fibre characteristically splits into two branches at the surface of the medulla, one ending in the lamina, one in the medulla.

The centrifugal fibres originate from all cell bodies in the medulla or even further proximally. They have either unbranching arms in the lamina or horizontally-running branches.

3. Discussion of the Connection Pattern

The retinula cell axons can be distinguished, as can the visual cells in the retina, by their diameter and position in the bundle. They form a regular pattern that scarcely varies from the proximal retina to the CBL: three thin axons lying on a straight line and perpendicular to these lie the two thickest fibres which have on both sides fibres of middle size.

Golgi analysis yields three types of fibres: two R(d) ending in EPL(B), four (R(s)1 and R(s)2) with different terminal structures in EPL(A) and three lvf reaching the medulla without branching. A correlation between pseudocartridge-elements and lamina terminals is possible considering their position in the axon bundles and their diameters the two thick fibres are R(d), the four middle ones R(s)1 and R(s)2, the three thin ones are lvf. (This model has been recently confirmed by Sommer and Wehner, in preparation). The four middle-size fibres belong to the cells 2, 3, 5 and 6 (Gribakin's green-receptors 1969, 1972). The thick fibres might be the axons of either cell 1 and 4 (Gribakin's UV-receptors) or cell 7 and 8 (Gribakin's blue-receptors).

As the 9nth retinula cell (UV-receptor according to Menzel and Snyder, 1974) always has a lvf we must assume that the other two lvf's belong to 7 and 8 or 1 and 4. This question is now being investigated.

According to the results of three different methods (methylene blue staining, reduced and selective silver-impregnation) three L-fibres can be attributed to each cartridge. It follows that six fibres leave the lamina and reach the medulla through the OCh.

Indirect evidence for the distribution of the fibre-types per cartridge may be drawn from the total number of fibre-types found. According to these all three retinula fibre-types should be uniformly distributed throughout the lamina. The same can be said of the L-fibres L(1), L(2) and L(3). The other fibres are less frequent.

The horizontal branching pattern of the different fibres in the different EPL-levels suggests synaptic contacts. If this is true, it can be said that the short visual cell axons have synaptic contacts with all L-fibres which themselves send collaterals to different EPL-levels depending on their type. The lateral elements of L(4) fibres in the EPL(C) have a maximum range of 80 μm in frontal sections. It is possible that they connect different cartridges (at a maximum distance of 5 cartridges) with each other.

Acknowledgements

I would like to thank my tutor Professor R. Wehner for his guidance and support of this paper and also Professor V. Braitenberg for enabling me to use his facilities at the Max-Planck-Institut für biologische Kybernetik in Tübingen, West Germany.

My special thanks to Dr. N. J. Strausfeld, Max-Planck-Institut für biologische Kybernetik, Tübingen, for his encouragement, fruitful discussions and friendly, unstinting help. I thank P. L. Herrling for helping me to prepare this paper in English.

The work was supported by grants No. 3.315.70 and 3.814.72 by the Fonds National Suisse de la Recherche Scientifique and the Sandoz Foundation.

References

Autrum, H. J., Zwehl, V. v.: Zur spektralen Empfindlichkeit einzelner Sehzellen der Drohne (*Apis mell.*). Z. vergl. Physiol. **46**, 8–12 (1962)

Autrum, H. J., Zwehl, V. v.: Die spektrale Empfindlichkeit einzelner Sehzellen des Bienenauges. Z. vergl. Physiol. **48**, 357–384 (1964)

Blest, A. D.: Some modification of Holme's silver nitrate method for insect central nervous system. Quart. J. micr. Sci. **102**, 413–417 (1961)

Bodian, D.: A new method for staining nerve endings in mounted paraffin sections. Anat. Rec. **65**, 89–97 (1936)

Braitenberg, V.: Unsymmetrische Projektion der Retinulazellen auf die Lamina ganglionaris bei der Fliege *Musca domestica*. Z. vergl. Physiol. **52**, 212–214 (1966)

Cajal, S. R.: Nota sobre la estructura de la mosca (*M. vomitoria* L.). Trab. Lab. Invest. Biol. Univ. Madrid **7**, 217–257 (1909)

Cajal, S. R., Sanchez, D.: Contribucion al conocimiento de los centros nerviosos de los insectos. Trab. Lab. Invest. Biol. Univ. Madrid **13**, 1–168 (1915)

Ciaccio, M. G.: L'oeil des Dipteres. J. Zool. (Paris) **5**, 313–319 (1876)

Colonnier, M.: The tangential organization of the visual cortex. J. Anat. (Lond.) **98**, 327–344 (1964)

Duelli, P., Wehner, R.: The spectral sensivity of polarized light orientation in *Cataglyphis bicolor* (Formicidae, Hymenoptera). J. comp. Physiol. **86**, 37–53 (1973)

Eguchi, E., Waterman, T. H.: Localization of the violet and yellow receptor cells in the crayfish retinula. J. gen. Physiol. **62**, 355–374 (1973)

Goldsmith, T. H.: Fine structure of the retinulae in the compound eye of the honey bee. J. Cell. Biol. **14**, 489–494 (1962)

Goldsmith, T. H.: The visual system of insects. The physiology of insects (ed. Rockstein). New York: Academic Press 1964

Gribakin, F. G.: Die Typen der photorezept. Zellen des zusammengesetzten Auges der Arbeiterin aufgrund der Elektronen-Mikroskopie. Zytologie. Band IX, Nr. 10 Moskau 1967

Gribakin, F. G.: Cellular basis of colour vision in the honey bee. Nature (Lond.) **223**, 639–641 (1969)

Gribakin, F. G.: The distribution of the long wave photoreceptors in the compound eye of the honey bee as revealed by selective osmic staining. Vision Res. 12, 1225–1230 (1972)

Grundler, O. J.: Elektronenmikr. Untersuchungen am Auge von *Apis mell*. Zulassungsarbeit zur wiss. Prüfung, Univ. Würzburg 1972

Grundler, O. J.: Morphologische Untersuchungen am Bienenauge nach Bestrahlung mit Licht versch. Wellenlängen. Cytobiologie, Band 7, Heft 1, 105–110 (1973)

Hickson, S. J.: The eye and optic tract of insects. Quart. J. micr. Sci. 2, Band XXV (1885)

Holmes, W.: Silver staining of nerve axons in paraffin sections. Anat. Rec. 86, 157–188 (1943)

Jonescu, C.: Vergleichende Untersuchungen über das Gehirn der Honigbiene. Jena. Z. Med. Naturw. 45 (N.F. 38), 111–180 (1909)

Karnovsky, M. J.: A formaldehyde-glutaraldehyde fixative of high osmolarity for use in electronmicroscopy. J. Cell. Biol. 27, 137 A (1965)

Kenyon, F. C.: The brain of the bee. A preliminary contribution to the morphology of the nervous system of the arthropoda. J. comp. Neurol. 6, 133–210 (1896)

Kenyon, F. C.: The optic lobes of the bees brain in the light of recent neurological methods. Amer. Nat. 31, 369–377 (1897)

Kirschfeld, K.: Die Projektion der optischen Umwelt auf das Raster der Rhabdomere im Komplexauge von *Musca*. Brain Res. 3, 248–270 (1967)

Leydig, F.: Zum feineren Bau der Arthropoden. Arch. Anat. Physiol. (1885)

Menzel, R., Snyder, A. W.: Polarized light detection in the bee, *Apis mellifera*. J. comp. Physiol. 88, 247–270 (1974)

Perrelet, A.: The fine structure of the retina of the honey bee drone: an electron microscopical study. Z. Zellforsch. 108, 530–562 (1970)

Perrelet, A., Baumann, F.: Evidence for extracellular space in the rhabdome of the honey bee drone eye. J. Cell. Biol. 40, 825–830 (1969a)

Perrelet, A., Baumann, F.: Presence of three small retinula cells in the ommatidium of the honey bee drone eye. J. Microscopie, 8 (No. 4), 497–502 (1969b)

Ribi, W. A.: Neurons in the first synaptic region of the bee, *Apis mell*. Cell Tissue Res. 148, 277–286 (1974)

Romeis, B.: Mikroskopische Technik. München: Oldenbourg 1948

Rothenbach, D.: The lamina of the honey bee. In prep.

Rowell, F. C. H.: A general method for silvering invertebrate central nervous systems. Quart. J. micr. Sci. 104, 81–87 (1963)

Skrzipek, K. H., Skrzipek, H.: Die Morphologie der Bienenretina (*Apis mellifica* L.) in elektronenmikroskopischer und lichtmikroskopischer Sicht. Z. Zellforsch. 119, 552–576 (1971)

Skrzipek, K. H., Skrzipek, H.: Die Anordnung der Ommatidien in der Retina der Biene (*Apis mell*.). Z. Zellforsch. 139, 567–582 (1973)

Snyder, A. W., Menzel, R., Laughlin, S. B.: Structure and function of the fused rhabdom. J. comp. Physiol. 87, 99–135 (1973)

Sommer, E. W., Wehner, R.: The 9nth retinula cell and the projection of visual cells of one Ommatidium to the first optic ganglion of the honey bee (*Apis mell*.). A Golgi-EM-study. In prep.

Strausfeld, N. J.: Variations and invariants of cell arrangements in the visual system and corpora ped.). Verh. zool. Ges. 64, 97–108 (1970)

Strausfeld, N. J.: The organization of the insect visual system (Light microscopy). I. Projections and arrangements of neurons in the lamina ganglionaris of Diptera. Z. Zellforsch. 121, 377–441 (1971)

Strausfeld, N. J., Blest, A. D.: Golgi studies on insects. Part. I. The optic lobes of Lepidoptera. Phil. Trans. Roy. Soc. London, B 258, 81–134 (1970)

Trujillo-Cenoz, O., Melamed, J.: On the fine structure of the photoreceptor second order neuron synapse in the insect retina. Z. Zellforsch. Abt. Histochem. 59, 71–77 (1963)

Varela, F. G.: Fine structure of the visual system of the honey bee (*Apis mellifica*). II. The lamina. J. Ultrastruct. Res. 31, 178–194 (1970)

Varela, F. G., Porter, K. R.: Fine structure of the visual system of the honey bee (*Apis mellifica*). J. Ultrastruct. Res. 29, 236–259 (1969)

Varela, F. G., Wiitanen, W.: The optics of the compound eye of the honey bee (*Apis mellifera*). J. gen. Physiol. 55, 336–358 (1970)

Weiss, M. J.: A reduced silver staining method applicable to dense neuropiles, neuroendocrine organs, and other structures in insects. Brain Res. 39, 268–273 (1972)

Subject Index

amacrine cells 28, 38, 40
axonbundle of one ommatidium, see pseudocartridge

basal membrane 10, 17

cartridge 17, 36, 37, 38
cell body layer 10, 13
centrifugal fibres 28, 38, 40
chiasma
— outer chiasma (externa) 17, 25, 28, 39
— inner chiasma (interna) 28

external plexiform layer 10, 14, 25, 38
e-vector sensitivities 36

fenestrated layer 10
first optic ganglion, see lamina ganglionaris
first synaptic region, see external plexiform layer
fly
— rotation of the R-fibre bundles 15, 36
fused rhabdom, see rhabdom

ganglion, see lamina ganglionaris, medulla, lobula
glial cells 14, 17, 37
Golgi's selective silver staining 9, 37

inserta sedis fibres 28

L-fibres 17, 21–25, 29, 32, 37, 38, 40, 41
lamina ganglionaris 10, 39
lobula 10
lobula plate 10
long visual fibres, see visual fibres

medulla (medulla externa) 10, 17, 25, 28–32, 39
methylene blue staining 8, 36
microvilli direction 14, 32, 36
monopolar cell bodies 10, 13, 37
monopolar cell fibres of the lamina, see L-fibres

Musca
— L-fibres 21

open rhabdom, see rhabdom
ommatidium 17, 36, 37
outer chiasma 10, 39

pigment cells 17, 33
pseudocartridge 10, 14, 20, 36
projection
— retina-lamina projection 14–17
— lamina-medulla projection 17, 28

R-fibres, see visual fibres
receptor cells 14, 32–36
reduced silver staining 8
retina 14, 32, 39
retinula cells, see receptor cells
retinula cell axon, see visual fibres
retinula cell axon bundle, see pseudocartridge
rhabdom
— fused rhabdom 14
— open rhabdom 15

second optic ganglion, see medulla
spectral sensitivity 32
short visual fibres, see visual fibres
synaptic contacts 20, 38

tangential fibres 17, 25–27, 38, 40
third optic ganglion, see lobula
trachea 10
tracheoblasts 14, 37
tracheoles 11

UV-sensitive receptor 33, 36

visual fibres 17–20, 32, 37–40
— short visual fibres 17, 38, 39
— — with deep ending (R(d)) 17, 38
— — — shallow ending (R(s)) 17, 18, 38
— long visual fibres 17, 20, 28, 38, 39

If you have any concerns about our products,
you can contact us on
ProductSafety@springernature.com

In case Publisher is established outside the EU,
the EU authorized representative is:
**Springer Nature Customer Service Center GmbH
Europaplatz 3, 69115 Heidelberg, Germany**

Printed by Libri Plureos GmbH
in Hamburg, Germany